FORTSCHRITTE DER PHYSIK

HERAUSGEGEBEN IM AUFTRAGE DER PHYSIKALISCHEN GESELLSCHAFT
DER
DEUTSCHEN DEMOKRATISCHEN REPUBLIK
VON F. KASCHLUHN, A. LÖSCHE, R. RITSCHL UND R. ROMPE

HEFT 9 · 1982 · BAND 30

AKADEMIE-VERLAG · BERLIN

ISSN 0015-8208 Fortschr. Phys., Berlin 30 (1982) 9, 451–505 EVP 10,– M 31728

BEZUGSMÖGLICHKEITEN

Bestellungen sind zu richten

- **in der DDR** an eine Buchhandlung
  oder an den AKADEMIE-VERLAG,
  DDR - 1086 Berlin, Leipziger Straße 3—4
- **im sozialistischen Ausland** an eine Buchhandlung für fremdsprachige Literatur oder an den zuständigen Postzeitungsvertrieb
- **in der BRD und Berlin (West)** an eine Buchhandlung oder an die Auslieferungsstelle
  KUNST UND WISSEN, Erich Bieber OHG,
  D - 7000 Stuttgart 1, Wilhelmstraße 4—6
- **in den übrigen westeuropäischen Ländern** an eine Buchhandlung oder an die Auslieferungsstelle
  KUNST UND WISSEN, Erich Bieber GmbH,
  CH - 8008 Zürich, Dufourstraße 51
- **im übrigen Ausland** an den Internationalen Buch- und Zeitschriftenhandel; den Buchexport, Volkseigener Außenhandelsbetrieb der Deutschen Demokratischen Republik,
  DDR - 7010 Leipzig, Postfach 160; oder

  an den AKADEMIE-VERLAG,
  DDR - 1086 Berlin, Leipziger Straße 3—4

Zeitschrift „Fortschritte der Physik"

Herausgeber: Prof. Dr. Frank Kaschluhn, Prof. Dr. Artur Lösche, Prof. Dr. Rudolf Ritschl, Prof. Dr. Robert Rompe, im Auftrag der Physikalischen Gesellschaft der Deutschen Demokratischen Republik.
Verlag: Akademie-Verlag, DDR - 1086 Berlin, Leipziger Straße 3—4; Fernruf: 22 36 221 und 22 36 229; Telex-Nr. 114420; Bank: Staatsbank der DDR, Berlin, Konto-Nr. 6836-26-20712.
Chefredakteur: Dr. Lutz Rothkirch.
Anschrift der Redaktion: Sektion Physik der Humboldt-Universität zu Berlin, DDR - 1040 Berlin, Hessische Straße 2.
Veröffentlicht unter der Lizenznummer 1324 des Presseamtes beim Vorsitzenden des Ministerrates der Deutschen Demokratischen Republik.
Gesamtherstellung: VEB Druckhaus „Maxim Gorki", DDR - 7400 Altenburg, Carl-von-Ossietzky-Straße 30/31.
Erscheinungsweise: Die Zeitschrift „Fortschritte der Physik" erscheint monatlich. Die 12 Hefte eines Jahres bilden einen Band. Bezugspreis je Band 180,— M zuzüglich Versandspesen (Preis für die DDR: 120,— M). Preis je Heft 15,— M (Preis für die DDR: 10,— M).
Bestellnummer dieses Heftes: 1027/30/9.

AN (EDV) 57618

Fortschritte der Physik 30, (1982) 9, 451—505

Technicolor, Extended-Technicolor and Tumbling

S. KAPTANOĞLU and NAMIK K. PAK

Middle East Technical University, Department of Physics, Ankara, Turkey

Contents

1. Dynamical Symmetry Breaking

a) A Review of the Standard Theory

Nowadays it is the common belief that the weak and electromagnetic interactions are governed by standard $SU(2) \otimes U(1)$ theory [1] which is based on the idea that some gauge symmetries are spontaneously broken.

For every spontaneously broken symmetry there is a Nambu-Goldstone boson (NGB) which may or may not be eliminated by the Higgs mechanism [2]. The detailed nature of the symmetry breaking (thus the dynamics of the theory) can be categorized according to the nature of NGB's. The NGB's may be elementary or composite. In the former case the Lagrangean will contain elementary scalar fields, and the spontaneous symmetry breaking (SSB) responsible for the intermediate vector-boson masses is due to the vacuum expectation values of these elementary scalars. In the latter case it is assumed that the NGB's associated with SSB are bound states; thus the symmetry breaking can be of purely dynamical nature.

To motivate ourselves for the latter alternative let us briefly review the main features of the standard theory. It is described by the Lagrangian

$$\mathcal{L} = -\frac{1}{4} W^a_{\mu\nu} W^{\mu\nu,a} + \bar{\psi} i \not{D} \psi + (D_\mu \varphi)^+ (D^\mu \varphi) - P(\varphi) \tag{1.1}$$

where $P(\varphi)$ is a real 4th order polynomical in φ. The electro-weak gauge symmetry, G_w, is broken by the vacuum expectation value of the Higgs fields. To lowest order, these vacuum expectation values are determined by the symmetry breaking condition

$$\left.\frac{\partial P(\varphi)}{\partial \varphi_i}\right|_{\varphi_i = v_i} = 0. \tag{1.2}$$

Denoting the generators of G_w by ϑ^α (including the coupling constants), this gives a zeroth order mass to the weakly interacting gauge bosons described by the mass matrix

$$(m_w{}^2)_{\alpha\beta} = v_i(\vartheta_\alpha \vartheta_\beta v)_i\,, \tag{1.3}$$

and to the Fermions described by the mass matrix

$$m_\psi = \Gamma_i v_i\,. \tag{1.4}$$

The magnitude of v is fixed by the observation that the Fermi weak-interaction constant is of order $e^2/m_w{}^2$. Since ϑ is of order e, this, then, gives $v \simeq m_w/e \simeq 1/\sqrt{G_F} \simeq 300$ GeV. An important consequence of the standard Higgs mechanism is the empirically established relation [3]

$$\frac{m_z}{m_w} \cos \theta_w = 1 + o(\alpha)\,. \tag{1.5}$$

One alarming feature of the standard theory is the appearance of a physical elementary scalar, the Higgs particle. There is no a priori way to determine its mass and self-coupling (or equivalently the Yukawa coupling, and the self-coupling) in the theory; thus they are totally arbitrary parameters to lowest order. However, higher order radiative corrections bring in some constraints [4]:

For instance it is impossible to make the scalar self-coupling, λ arbitrarily small, because it receives corrections of the order α^2 (this, then, implies for instance that $m_H \geqq 5$ GeV). Still one can adjust λ to reasonably small values.

Radiative corrections to the Yukawa Coupling, Γ, don't receive contribution of the order α^2, but only $G_F{}^2$. Thus, Yukawa coupling can be chosen to be arbitrarily small to have, say, the electron mass much smaller than the gauge boson mass.

The standard model with its arbitrary parameters adjusted such that $\lambda \ll 1$, $\Gamma \ll 1$, seems to be so successfull phenomenologically that it is in order to ask why we look for other alternatives, which pays the price of introducing new short strong of interactions as the driving force which break the electro-weak symmetry.

There are several reasons which are tabulated below. However we have to emphasize that these problems don't impeach the consistency of the standard theory as a phenomenological description, but they do indicate that this description is no more than a phenomenology:

1. The first problem is arbitrariness. In order to define a realistic theory it is necessary to specifiy the values of a large number of parameters. Each of these parameters may be freely adjusted. These parameters include the gauge couplings, g and g', for the $SU(2)_L \otimes U(1)_Y$ case, the scalar self-couplings, and the Yukawa Couplings of Higgses to Fermions. None of these parameters can be predicted by the theory.
A severer problem than finding a large number of adjustable parameters is that some of these parameters must be adjusted to extremely small values. For example the observed value of the electron mass requires that $\Gamma_e \simeq 10^{-6}$.

2. Gauge Hierarchy problem [5]. This problem arises when one attempts to find a remedy to the first problem: Because $SU(2) \otimes U(1)$ is not simple, there are two free gauge coupling constants, and we have a priori no way to determine the mixing angle (problem becomes more acute when we try to unify strong interactions with the electroweak interactions). In order to reduce the arbitrariness we may suppose that $SU(2) \otimes U(1)$ is a subgroup of a larger simple gauge group G. But we do not observe weak interactions corresponding to the generators of $\mathcal{G}$ outside the $SU(2) \otimes U(1)$ subalgebra. Therefore, since there is supposed to be only one gauge coupling constant, we must *assume* that the gauge bosons of $G/SU(2) \otimes U(1)$ are much heavier than W and Z. This can only happen if the breaking $G \to SU(2) \otimes U(1)$ is much stronger than the breaking $SU(2) \otimes U(1) \to U(1)_\gamma$. All the previous attempts to solve this problem in the standard model have failed. We shall shortly see that models based on the idea of dynamical symmetry breaking may naturally lead to a hierarchical breakdown of gauge symmetries.

3. Problem of fine tuning — Violation of Naturalness: Effective interactions at a large length scale (low energy) should follow from the properties at a much smaller fundamental length scale (large energy) $\varkappa^{-1}$, which serves as a real cutoff ($\varkappa$ can be taken to be of the order 10^{19} GeV, corresponding to the Planck gravitational length). The basic parameters of such a theory are a set of dimensionless base coupling constants, g_0 and base masses, m_0. The existence of a light mass spectrum of 0(1 GeV) implies that the physical properties at low energy to be stable against very small variation of g_0 and μ_0 ($\mu_0 \equiv m_0/\varkappa$). Given that the light spectrum has mass 10^{19} times smaller than that of the fundamental scale, one is forced to ask what kind of special adjustments of parameters must be made in order to insure such gigantic ratios of mass scales.

That elementary scalar fields receive a self-energy quadratic in $\varkappa$ [4] put them in conflict with the principle of naturalness; because this leads to an extremely unnatural adjustment of the bare parameters:

$$m^2 = m_0^2 + \Delta m^2 = m_0^2 + \varkappa^2 g_0^2. \tag{1.6}$$

Solving for μ_0^2, we get

$$\mu_0^2 = \frac{m_0^2}{\varkappa^2} = \frac{m^2}{\varkappa^2} - g_0^2 = -g_0^2\left[1 - \frac{1}{g_0^2}\left(\frac{m^2}{\varkappa^2}\right)\right]. \tag{1.7}$$

The result $m^2/\varkappa^2 \simeq 10^{-38}$ shows that μ_0^2 should be adjusted to the 38th decimal place. If it is not, then the physical masses would come out to be of the order 10^{19} GeV.
Unfortunately standard theory, containing elementary scalar fields, cannot avoid such unnatural adjustments, because of the quadratic divergences in the scalar particle masses.
But fortunately there is a class of theories where enormous mass ratios occur naturally — the asymptotically free (AF) theories [6].

In asymptotically-free theories the running coupling-constant satisfies

$$m \frac{\partial g(m)}{\partial m} = -Cm^3 + o(m^5), \tag{1.8}$$

where m is the momentum scale at which the coupling is measured and C is a group theoretical constant. For AF theories $C > 0$, hence g increases towards the infrared region. Integration of the above equation gives:

$$\frac{1}{g^2(m)} = \frac{1}{g^2(\varkappa)} + C \log \frac{m^2}{\varkappa^2}. \tag{1.9}$$

We shall identify $g^2(\varkappa)$ with the bare coupling g_0. The value of m where g becomes large (where this perturbative formula doesn't hold anymore) is given by

$$\frac{m}{\varkappa} \simeq \exp\left\{-\frac{1}{2Cg_0{}^2}\right\}, \tag{1.10}$$

For pure $SU(3)$ Yang-Mills theory, $C = 11/16\pi^2$. Thus to make $m/\varkappa \simeq 10^{-19}$, we need to have $g_0{}^2 \simeq 0.2$, a natural value.

Thus we see that the asymptotically free alternatives, which don't contain elementary scalar fields could free the standard theory from these difficulties. A promising alternative is based on the idea of Dynamical Symmetry Breaking (DSB) [7].

b) SSB with Composite Goldstone Bosons — Dynamical Symmetry Breaking

In this new approach the electro-weak breaking is accomplished by *a new strong* interaction of fermions and gauge bosons. The Higgs bosons are replaced by the composites (bound states) of new types of fermions.

This program is simply to build up the Higgs mesons in just the same way that the hadrons are built in QCD from fermions of zero bare mass, bound by a strongly coupled gauge field. Such a theory will have a rich structure, but it will introduce only a single parameter, for each new strong interaction gauge group (a mass scale Λ, where the corresponding running coupling constant becomes strong). In the best case there should be only one such parameter. For instance, if we make the approximation of ignoring the bare masses of the u, d quarks (and the influence of heavy flavors) QCD has only one adjustable parameter, the gauge field coupling constant g_c. But g_c varies with the momentum transfer involved in a particular process, growing small [6] at large momenta, and vice versa. This means that what we actually have the freedom to adjust is the value of g_c at a certain momentum, or better at the momentum Λ where g_c reaches a certain fixed (large) value. Since this momentum Λ is the only adjustable parameter in the theory, all the other quantities appearing in the theory are in principle computable in terms of it. One further nice feature of QCD-like, asymptotically free theories is that enormous mass scale ratios may occur naturally as was pointed out above.

Let us briefly review the motivation for introducing a new type of strong interaction rather than utilizing the good old strong color interactions for the purpose of breaking $SU(2) \otimes U(1)$ symmetry. For this, consider ordinary QCD with just a massless doublet, coupled strongly to color and weakly to gauge bosons. The standard scenario in DSB contains two stages: [8]

a) First turn off the electro-weak couplings. Then look what structure develops from the pure strong color theory.

b) Then add the electro-weak interactions as a perturbation, and see how they do effect the structure.

In a QCD-like gauge theory the coupling of a gauge boson to fermions preserves the fermion helicity:

$$\mathcal{L} = \bar{\psi}(i\not{D} - m)\,\psi - \frac{1}{4}\,F_{\mu\nu}^2 = \bar{\psi}_L i\not{D}\psi_L + \bar{\psi}_R i\not{D}\psi_R - m(\bar{\psi}_L\psi_R + \bar{\psi}_R\psi_L) - \frac{1}{4}\,F_{\mu\nu}^2 \tag{1.11}$$

where

$$\psi_L \equiv \frac{1-\gamma_5}{2}\,\psi, \qquad \psi_R \equiv \frac{1+\gamma_5}{2}\,\psi. \tag{1.12}$$

To these symmetries would be associated the following conserved currents:

$$\begin{aligned} J^a_{\mu_L} &= \bar{\psi}\gamma_\mu\frac{\tau_a}{2}\left(\frac{1-\gamma_5}{2}\right)\psi = \bar{\psi}_L\gamma_\mu\frac{\tau^a}{2}\,\psi_L \\ J^a_{\mu_R} &= \bar{\psi}\gamma_\mu\frac{\tau_a}{2}\left(\frac{1+\gamma_5}{2}\right)\psi = \bar{\psi}_R\gamma_\mu\frac{\tau^a}{2}\,\psi_R. \end{aligned} \tag{1.13}$$

The group of these handed flavor symmetries, $SU(2)_L \otimes SU(2)_R$ is referred as the chiral group.

The key idea of DSB is, when the gauge interaction become sufficiently strong, the fermion bound state scalars condensate, and the vacuum is filled with an infinite number of $\bar{\psi}\psi$ pairs (similar to the pair condensates of superconductivity), as well as condensates of the form $\bar{\psi}\psi\bar{\psi}\psi$, etc. ...

Let us look at such a pair. This must have the vacuum quantum numbers (zero momentum and angular momentum)

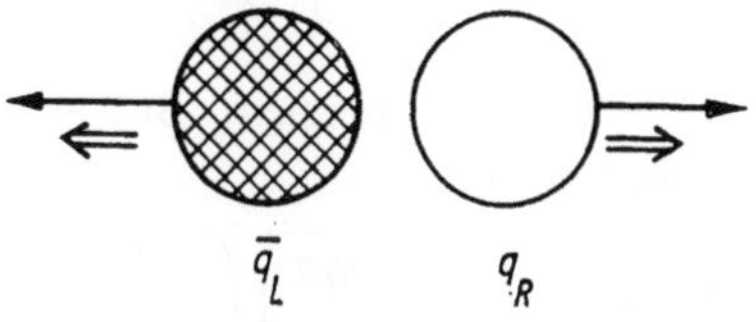

Fig. 1.1

Thus two separate $SU(2)$ groups cannot be respected anymore; they are linked by pairing: q_R has been forced to pair with $\bar{q}_L$, and vice versa. Thus the ordinary isospin

$$J_\mu{}^a = J^a_{\mu_L} + J^a_{\mu_R} = \bar{\psi}\gamma_\mu\frac{\tau_a}{2}\,\psi \tag{1.14}$$

is preserved, but

$$J^a_{\mu_5} = -J^a_{\mu_L} + J^a_{\mu_R} = \bar{\psi}\gamma_\mu\gamma_5\,\frac{\tau_a}{2}\,\psi \tag{1.15}$$

is (spontaneously) broken. Since the vacuum does not respect the separate symmetries (L, R), the u, d quarks can acquire dynamical masses through their interaction with the condensates.

Recalling that to every spontaneously broken generator there corresponds a massless pseudoscalar particle, NGB, the spontaneous breaking of $J^a_{\mu_5}$ yields a triplet of 0^- particles:

$$\langle 0| J^a_{\mu_5} |\pi^b(p)\rangle = ip_\mu F_\pi \delta^{ab}. \tag{1.16}$$

Now, that happens when we couple this theory (with its dynamically broken state) to electro-weak (EW) interactions? The crucial feature is that the EW interactions involve handed currents, and not simply $J_\mu{}^a$. This means that the gauge bosons couple to the pair-condensate which can communicate to them its symmetry breaking, and as a result, give, the gauge bosons masses. This brings us to the subject of mass generation in DSB scheme.

It is desirable, but very difficult to compute the masses generated for weak gauge bosons by straightforward examination of the condensate, but it can be done easily using the indirect method of SCHWINGER [7]. If a vector boson is to acquire a mass m^2, its self energy (vacuum polarization) $\Pi_{\mu\nu}(q^2)$ must tend to m^2 as $q \to 0$. Indeed writing the vacuum polarization tensor of the gauge boson as (it is transverse because of gauge invariance)

$$\Pi_{\mu\nu}(q) = (q^2 g_{\mu\nu} - q_\mu q_\nu)\, \Pi(q^2) \tag{1.17}$$

the gauge-boson propagator is

$$D_{\mu\nu}(q) = \frac{(q_{\mu\nu} - q_\mu q_\nu/q^2)}{q^2(1 - \Pi(q^2))} + \text{(gauge dependent terms)} \tag{1.18}$$

we see that if $\Pi(q^2)$ has a pole at $q^2 = 0$, then the pole at $q^2 = 0$ in $D_{\mu\nu}(q^2)$ is replaced by a pole at some finite mass m^2, which is given by the residue of the pole in $\Pi(q^2)$. Indeed writing $\Pi(q^2)$ as

$$\Pi(q^2) = \frac{\beta}{q^2 - q_0{}^2} + \text{(analytic terms)} \tag{1.19}$$

then the propagator is

$$D_{\mu\nu}(q^2) = \left(q_{\mu\nu} - \frac{q_\mu q_\nu}{q^2}\right) \frac{1}{q^2} \frac{q^2 - q_0{}^2}{q^2 - (\beta + q_0{}^2)}. \tag{1.20}$$

If $q_0{}^2 \neq 0$, then the pole at $q^2 = 0$ of $D_{\mu\nu}$ persists. But as $q_0 \to 0$, i.e., if $\Pi(q^2) \underset{q\to 0}{\simeq} \beta/q^2$ then the pole structure of $D_{\mu\nu}$ becomes

$$D_{\mu\nu}(q^2) = \left(g_{\mu\nu} - \frac{q_\mu q_\nu}{q^2}\right) \left(\frac{1}{q^2 - \beta}\right). \tag{1.21}$$

Now the question is how could such a pole in $\Pi(q^2)$ arise? If there exist NGB's in the theory coupled to some dynamically broken generators of a chiral symmetry, then the direct NGB-Vector Boson (VB) coupling could give such a contribution to the vacuum polarization:

We first denote by $j^a_{\mu_L}$ the purely left-handed electro-weak current, coupled to the gauge boson whose mass we are computing. The coupling of its dynamically broken part is obtained from

$$g j^a_{\mu_L} W_\mu{}^a = \frac{g}{2} J_\mu{}^a W_\mu{}^a - \frac{g}{2} J^a_{\mu_5} W_\mu{}^a. \tag{1.22}$$

This, combined with (1.17), gives us the NGB-VB coupling as $iq_\mu F_\pi g/2$. Thus the vacuum polarization tensor is

$$\Pi_{\mu\nu} = -\frac{q_\mu q_\nu}{q^2} \left(\frac{g^2 F_\pi{}^2}{4}\right). \tag{1.23}$$

Due to the current conservation, this would have to be supplemented by a $g_{\mu\nu}$-piece, yielding

$$\Pi_{\mu\nu}(q^2) = \left(g_{\mu\nu} - \frac{q_\mu q_\nu}{q^2}\right)\left(\frac{g^2 F_\pi^2}{4}\right). \tag{1.24}$$

Thus, we see that those gauge bosons, coupled to the dynamically broken symmetry generators acquire a mass:

$$m_w^2 = \frac{1}{4} g^2 F_\pi^2. \tag{1.25}$$

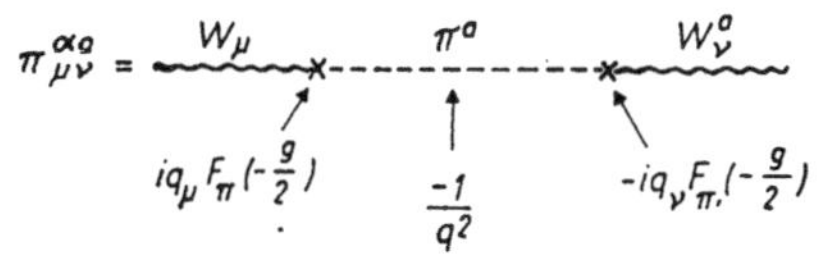

Fig. 1.2

c) Technicolor

(1.25) tells us that if we were to use the NGB's of the ordinary QCD, the pions, to generate the masses of the electro-weak gauge bosons, they would come out to be

$$m_w = \frac{1}{2} g f_\pi \simeq 30 \text{ MeV} \tag{1.26}$$

which is off by a factor of 10^3. Actually that ordinary QCD fails to give masses to $W^\pm$, Z via DSB is a very gratifying result. Because if this was not the case we would have a lot of trouble in explaining why we have pions as physical particles in the lower end of the spectrum.

Nevertheless the above neat idea can still be used: All that is needed is to suppose that there exists some other world not too dissimilar from the pion world (ordinary QCD), but on a mass scale sufficiently large to account for an intermediate boson mass, $m_w \simeq g/\sqrt{G_F} \simeq 70$ GeV. An indispensible element of this world is that in the limit of vanishing weak gauge coupling constant, there should exist at least a triplet of NGB's. The mass scale associated with this world can be estimated easily:

$$\frac{g}{2} F_\pi \simeq \frac{g}{\sqrt{G_F}} \Rightarrow F_\pi \simeq \frac{1}{\sqrt{G_F}} \simeq 300 \text{ GeV}. \tag{1.27}$$

Thus

$$\frac{F_\pi}{f_\pi} \simeq \frac{300 \text{ GeV}}{95 \text{ MeV}} \simeq 3000. \tag{1.28}$$

Since all the dimensional parameters in a dynamically broken theory are of order of Λ, where the strength of that particular interaction becomes $0(1)$, then the new type of strong interaction must be a scaled up analog of QCD with the characteristic scale

$$\Lambda_{TC} \simeq (300 \text{ GeV} - 1 \text{ TeV}). \tag{1.29}$$

Now consider a world with both QCD and colored quarks along with this new type QCD-like interactions, which we call Technicolor (TC) [9], and technifermions weakly coupled to the electro-weak gauge bosons. In this combined picture the linear combination,

$$|\text{pions absorbed}\rangle = \frac{F_\pi}{\sqrt{F_\pi^2 + f_\pi^2}} |\Pi\rangle + \frac{f_\pi}{\sqrt{F_\pi^2 + f_\pi^2}} |\pi\rangle \tag{1.30}$$

are eaten up by the electro-weak gauge bosons, while the orthogonal combinations

$$|\text{physical pions}\rangle = \frac{F_\pi}{\sqrt{F_\pi^2 + f_\pi^2}} |\pi\rangle - \frac{f_\pi}{\sqrt{F_\pi^2 + f_\pi^2}} |\Pi\rangle \tag{1.31}$$

remain as massless pions in the spectrum. Since $F_\pi \gg f_\pi$, the physical pion is mostly the QCD pion while the absorbed pion is mostly technipion.
To see how these particular combinations arise we first note that

$$\begin{aligned} \langle 0| J^a_{\mu_5} |\pi^a(q)\rangle &= i f_\pi q_\mu \\ \langle 0| J^a_{\mu_5} |\Pi^a(q)\rangle &= i F_\pi q_\mu \end{aligned} \tag{1.32}$$

where J_μ^5 is the full axial vector currents (involving both quarks and techniquarks). These give

$$\begin{aligned} \langle 0| J^a_{\mu_5} |\text{pion absorbed}\rangle &= \sqrt{F_\pi^2 + f_\pi^2}\, q_\mu \\ \langle 0| J^a_{\mu_5} |\text{physical pion}\rangle &= 0. \end{aligned} \tag{1.33}$$

The electro-weak gauge bosons couple to the pions through these axial currents so the physical pion has no coupling while the orthogonal combination is totally absorbed. In the standard theory also the same mixing occurs; for example the physical pion has a very small Higgs admixture.
There are no direct effects of G_{TC} interactions at ordinary energies if its chiral symmetry is totally broken dynamically. In this case all the TC-sector get dynamical mass of the order $\Lambda_{TC} \approx 1$ TeV. To illustrate the mass generation mechanism for the electro-weak gauge bosons we shall first consider a simple model in which there is no coupling between the TC-sector, and ordinary quarks and leptons.
The Lagrangean for our toy model is

$$\begin{aligned} \mathcal{L} = &-\frac{1}{4} F^A_{\mu\nu} F^{\mu\nu,A} - \frac{1}{4} A^a_{\mu\nu} A^{\mu\nu,a} - \frac{1}{4} B_{\mu\nu} B^{\mu\nu} \\ &+ \bar{Q} i\gamma_\mu [\partial_\mu - i g_{TC} \hat{F}_\mu - i g \hat{A}_\mu - i g' \hat{B}_\mu] Q \end{aligned} \tag{1.34}$$

where Q is a massless flavor doublet (UD) of techni-quarks (technicolor indices are suppressed as they are irrelevant for the following computation), and $\hat{F}_\mu$, $\hat{A}_\mu$, and $\hat{B}_\mu$ are constructed from the G_{TC}, $SU(2)_L$ and $U(1)_Y$ vector potentials, Dirac matrices, TC matrices (which we don't specify explicitly), and 2×2 flavor matrices. Denoting the average electric charge of the doublet by q, we have:

$$\begin{aligned} \hat{A}_\mu &= A_\mu{}^a \frac{\tau^a}{2} \frac{1 - \gamma_5}{2} \\ \hat{B}_\mu &= \left[\frac{1 + \gamma_5}{2} + 2q\right] B_\mu . \end{aligned} \tag{1.35}$$

Thus, leaving aside the pure TC-interaction part, the electro-weak interactions of the TC-quarks are governed by:

$$\mathcal{L}_w = \bar{Q}\gamma^\mu \left[g A_\mu{}^a \frac{\tau^a}{2} \left(\frac{1-\gamma_5}{2} \right) + g' B_\mu \left(\frac{\tau_3}{2} \left(\frac{1+\gamma_5}{2} \right) + 2q \right) \right] Q. \tag{1.36}$$

We can have one further simplification for the class of doublets with zero average charge, $q = 0$; this simply means that there is no vector piece in the B_μ-interactions. Thus

$$\mathcal{L}_w = \bar{Q}\gamma^\mu \left[g A_\mu{}^a \frac{\tau^a}{2} \left(\frac{1-\gamma_5}{2} \right) + g' B_\mu \frac{\tau_3}{2} \left(\frac{1+\gamma_5}{2} \right) \right] Q. \tag{1.37}$$

Now let us apply our standard scenario. In the limit where the electro-weak interactions are turned off, suppose that the chiral symmetry of the TC-sector, $SU(2)_L \otimes SZ(2)_R$, undergoes a DSB, yielding 3 NGB's:

$$SU(2)_L \otimes SU(2)_R \xrightarrow{\text{DSB}} SU(2)_V$$
$$\langle 0 | J^a_{\mu_5} | \pi^a(q) \rangle = i F_\pi q_\mu. \tag{1.38}$$

Then, the unbroken and the broken currents are

$$J_\mu{}^a = \bar{Q}\gamma_\mu \frac{\tau_a}{2} Q$$
$$J^a_{\mu_5} = \bar{Q}\gamma_\mu\gamma_5 \frac{\tau_a}{2} Q. \tag{1.39}$$

We can express the handed-currents entering into the weak-interactions of techniquarks in terms of these; thus the electro-weak interaction term can be rewritten as

$$\mathcal{L}_w = \frac{g}{2} A_\mu{}^a (J_\mu{}^a - J^a_{\mu_5}) + \frac{g'}{2} B_\mu (J_\mu{}^3 + J^3_{\mu_5}). \tag{1.40}$$

Now we can start computing the $q_\mu q_\nu$-term for the vacuum-polarization-tensor of the $A_\mu{}^a$ and B_μ vector bosons.
Thus electro-weak vector boson mass matrix is

$$M^2 = \frac{F_\pi{}^2}{4} \begin{pmatrix} g^2 & & & \\ & g^2 & & \\ & & g^2 & -gg' \\ & & -gg' & g'^2 \end{pmatrix}. \tag{1.41}$$

The eigenstates of the mass matrix can be found easily by diagonalizing it:

$$M_d{}^2 = \frac{F_\pi{}^2}{4} \begin{pmatrix} g^2 & & & \\ & g^2 & & \\ & & g^2 + g'^2 & \\ & & & 0 \end{pmatrix} \begin{matrix} A^1 \\ A^2 \\ Z \\ A \end{matrix} \tag{1.42}$$

and the neutral vector bosons with well-defined mass are:

$$Z_\mu = \frac{g A_\mu{}^3 - g' B_\mu}{\sqrt{g^2 + g'^2}} \qquad A_\mu = \frac{g' A_\mu{}^3 + g B_\mu}{\sqrt{g^2 + g'^2}}. \tag{1.43}$$

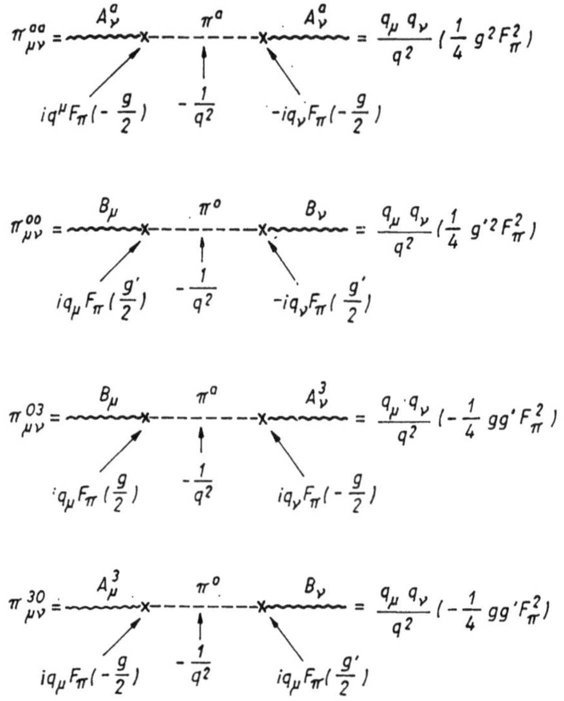

Fig. 1.3

Notice that the famous relation

$$\frac{M_z}{M_w} = \sqrt{1 + \frac{g'^2}{g^2}} = \frac{1}{\cos \theta_w} \tag{1.44}$$

is obtained automatically. It is remarkable that this relation holds, not only at the tree-level but to all orders of technicolor interactions (as they should, because these are extremely strong interactions) if in the particular model we are considering there exists a global $SU(2)$ symmetry group under which the generators of $SU(2)_L$ transform as a triplet, which is left unbroken by the technicolor interactions [*10*]. This $SU(2)$ group is given the name Custodial-$SU(2)$ in honor of its protective nature. The custodial $SU(2)$ can, but need not, be isospin. Of course there is another phenomenological fact, $M_u \neq M_d$, which need be made compatible with (1.44). It is interesting to note that $M_u \neq M_d$ signals a clear violation of isospin, whereas isospin conservation would imply (1.44).
The superiority of Technicolor scheme over the Higgs version is that almost always there exists a custodial $SU(2)$ in any realistic model. In the Higgs version this result is extremely representation dependent; any representation for the Higgs other than the doublet fails to produce (1.44).
Since we have started with a single doublet of techniquarks, there were 3 NGB's due to DSB of $G_f = SU(2)_L \otimes SU(2)_R$ which eventually are eaten up by the electro-weak vector gauge bosons. Thus no trace of this new TC-interaction is left. But as we shall see below the situation changes drastically as the number of techni-quark doublets increases. For instance if we have N doublets, then the chiral flavor symmetry would be $SU(2N)_L \otimes SU(2N)_R$. If this undergoes, say, the DSB,

$$SU(2N)_L \otimes SU(2N)_R \xrightarrow{\text{DSB}} SU(2N)_V,$$

then there emerges $4N^2 - 1$ NGB's, out of which again 3 are eaten by weak gauge bosons. Then we would have been left with $4N^2 - 4$ extra NGB's. Some of these extra NGB's will gain mass when we turn on the electro-weak perturbation. It is clear that a realistic model shouldn't have any light NGB's left in the end. Such (realistic) enlarged models are complex and they involve a very subtle problem which is called the "Vacuum Alignment" problem[1]); this will be the subject of the next section [*8*, *12*].

2. Vacuum Alignment

a) General Formulation of the Problem

Consider a set of massless techni-fermions (N of them) coupled strongly to gauge group G_{TC}. In the limit where weak interactions are turned off, the TC-interactions are necessarily invariant, not only under the strong gauge group G_{TC}, but also under a global techni-flavor group

$$G_f = SU(N)_L \otimes SU(N)_R .$$

When the running coupling constant of G_{TC} becomes strong, this strong interaction causes he G_f symmetry to break down to some subgroup H_f, by dynamical effect which allow some of the fermions coupled to G_{TC} to acquire a dynamical mass.
The DSB, $G \to H_f$ produces a manifold of vacuum states, one for each point in G_f/H_f (i.e. the vacua can be parametrized by the coset space), degenerate in energy before we turn on électro-weak interactions. A particular vacuum of the broken symmetry, $|0\rangle$, corresponds to a particular orientation of the subgroup H_f inside G_f. Now let us turn on the electro-weak interactions. Since G_w is a group of symmetries it must be a subgroup of G_f. Therefore it determines a second orientation inside G_f. The energy of a given state $|0\rangle$ (which is due to the explicit breaking by the electro-weak perturbation) will now depend on the angle between these two different orientations. This energy lifts the degeneracy of the states $\{|0\rangle\}$ and forces us to choose a particular one as the true vacuum. Then the problem is to find this true vacuum [*12*] (a perturbation expansion should be performed about the correct vacuum):

It is well known that there exists a NGB for every independent broken generator in G_f (those generators of G_f/H_f). However those NGB's which correspond to the generators

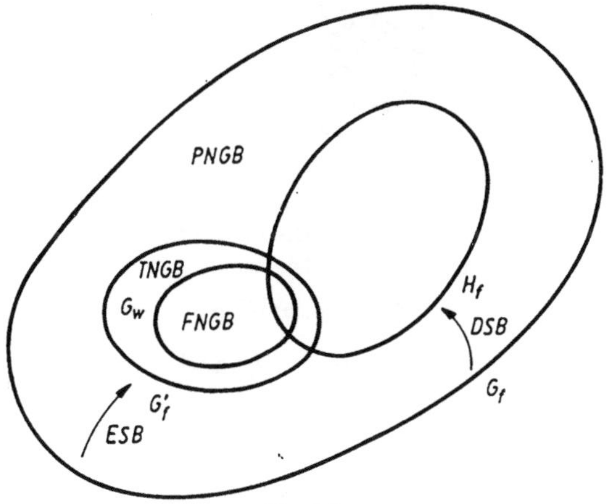

Fig. 2.1

[1]) Originally this dates back do DASHEN [*11*].

of G_w are eliminated by the Higgs mechanism; these are called fictitious Nambu-Goldstone bosons (FNGB). The other NGB's which are not eliminated by the Higgs mechanism (those which are neither in H_f nor G_w) gain masses of order $e\Lambda_{TC}$ due to electroweak perturbation. Weinberg has observed [8] that global symmetry group of the weak interactions, G_f', may be larger than the local gauge group G_w. Thus there may be left also true Nambu-Goldstone bosons (TNGB's) which correspond to the dynamically broken symmetries in G_f' not in G_w.

The vacuum alignment problem is a difficult problem for complicated group structures. Of course we have to make sure that there is a massless gauge boson, photon, left in the model at the end of the day as a consistency check. Shortly we shall obtain a strikingly interesting theorem which governs the vacuum alignment problem: The favored vacuum state is simply that one which minimizes the masses acquired by the G_w vector bosons. This is called "resistance to breaking".

To formulate the problem let us observe that the strong interactions alone do not determine which subgroup of G_f is left unbroken. Given any solution of the strong interaction dynamics with a subgroup H left invariant, we can find another subgroup left invariant as

$$H(g) = gHg^{-1} \tag{2.1}$$

where g is any element of G_f. Normally we would not be bothered by this ambiguity, if it were not for the electro-weak interactions which explicitly break G_f down to G_f'. Then the different solutions corresponding to different unbroken subgroups are physically inequivalent.

The lowest order electro-weak perturbation ΔH_w is generated by a G_w-boson exchange. The vacuum energy is given by

$$\Delta\varepsilon(g) = \langle 0, g| \, \Delta H_w \, |0, g\rangle \tag{2.2}$$

where $|0, g\rangle$ is the vacuum corresponding to the solution with $H(g) = gHg^{-1}$ as the unbroken subgroup. "g" that defines the "correct" solution in the presence of perturbation ΔH_w is defined by the condition that $\Delta\varepsilon(g)$ be a minimum.

Fig. 2.2

It may prove to be more convenient to keep the solution of the DSB fixed and vary the electro-weak gauge group is embedded into G_f. That is we fix the vacuum, and the unbroken subgroup H_f, instead let the electro-weak gauge group be

$$G_w(g) = g^{-1}G_wg . \tag{2.3}$$

This corresponds fixing the embedding $H_f \subset G_f$, and varying the orientation of $G(g)_w \subset G_f$ until the minimum vacuum energy is obtained:

$$\Delta\varepsilon(U) = \langle 0| \, U(g)^{-1} \, \Delta H U(g) \, |0\rangle = \langle 0| \, \Delta H(g) \, |0\rangle . \tag{2.4}$$

Here U is the unitary matrix representing an element of G.

Let us parametrize $U(g)$ in terms of chiral charges:

$$U(g) = \exp\,(iw^aQ^a) . \tag{2.5}$$

Thus

$$\Delta\varepsilon(U) = \langle 0| \, \Delta H + iw^a[\Delta H, Q^a] - \frac{1}{2} w^a w^b [[\Delta H, Q^a], Q^b] + \cdots |0\rangle . \tag{2.6}$$

So the condition that $\Delta\varepsilon(U)$ be stationary with respect to variations of g is equivalent to the condition that

$$\frac{d}{dw^a} \Delta\varepsilon(U)\Big|_{w^a=0} = 0 \tag{2.7}$$

which yields

$$\langle 0| \, [\Delta H, Q^a] \, |0\rangle = 0. \tag{2.8}$$

Because if $Q_a \in G/H$, then they create NGB's from vacuum, then this condition graphically states that the tadpole graphs in which a single NGB disappears or vice versa) into the vacuum necessarily vanish (of course otherwise NGB's can be produced spontaneously and the vacuum is unstable-thus the perturbation theory would break down).

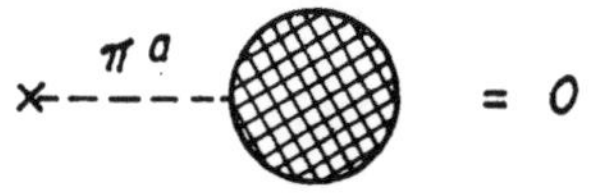

Fig. 2.3

Not only must we choose the gauge group $G_w(g)$ so that $\Delta\varepsilon(U)$ is stationary, but we must choose it so that $\Delta\varepsilon(U)$ is at least a local minimum. Again this condition is equivalent to

$$\frac{d^2}{dw^a \, dw^b} \Delta\varepsilon(U)\Big|_{w^a=0} = 0 \tag{2.9}$$

which yields

$$-\langle 0| \, [[\Delta H, Q^a], Q^b] \, |0\rangle \geqq 0. \tag{2.10}$$

Recognizing this quantity as the mass matrix of PNGB's [*13*], we thus see that this condition ensures the positivity of the mass matrix (a tachyonic NGB signals instability of the vacuum $|0\rangle$).

$$M^2_{ab} = -\frac{1}{F_a F_b} \langle 0| \, [[\Delta H, Q^a], Q^b] \, |0\rangle \geqq 0. \tag{2.11}$$

Before we make any attempt to use there formal expressions to obtain explicit results for various models let us make a brief detour and study the nature of the technifermionic condensate which supposedly affects the DSB.

b) Condensate, and the Symmetric Space Assumption

One typical signal of DSB of a global symmetry is the appearance of a non-zero vacuum expectation value for the operator $\bar{\psi}\psi$. In finding the detailed structure of this condensate we shall use QCD as a guide-line again: It so happens that in QCD with one doublet, the chiral flavor symmetry $SU(2)_L \otimes SU(2)_R \otimes U(1)$ is dynamically broken down to $SU(2)_V \otimes U(1)$. This DSB could be produced by a di-fermion condensate which is a color singlet, and which allows maximum flavor preservation any other di-fermion condensate will break the symmetry to a smaller subgroup than $SU(2)_V \otimes U(1)$.

Now we would like to simulate this with all possible DSB patterns. To do this we shall rewrite all techni-fermion fields as left-handed (LH) objects, by using two component

Weyl spinors. Thus $\psi_\alpha{}^{ai}$ denote the fermion fields; $\alpha = 1, 2$ is the spinor index, $a = 1, \ldots, m$ is the gauge group, G_S, index, and $i = 1, \ldots, N$ is the flavor index. Since we are going to be looking at the di-fermion condensates which are G_S singlets, it follows that the fermions must transform according to a real or a pseudo-real representation of G_S. Depending on the type of this representation the symmetry breaking pattern will be different. In general any real or pseudo-real representation must be a direct sum of any number of irreducible real and pseudo-real components and any number of complex representation along with their complex conjugates. In the light of these observations we see that there exist 4 basic types of models in which such a gauge-invariant chiral flavor symmetry breaking form can be constructed.

Case 1: ψ transforms under G_S according to the real representation $r \oplus \bar{r}$, where r is a complex irreducible representation (IRR) of G_S. Now we have $2N$ multiplets of 2-component Weyl spinors in hand.

$$\psi^{(r)ai} \in r(G_S) \qquad \psi^{(\bar{r})ai} \in \bar{r}(G_S). \tag{2.12}$$

Thus the chiral group is $G_f = SU(N) \times SU(N) \times U(1)$ and the general form of the condensate is

$$\Sigma = \langle \varepsilon_{\alpha\beta} \psi_\alpha{}^{(r)ai} \psi_\beta{}^{(\bar{r})bj} I_{ab,ij} \rangle \neq 0 \tag{2.13}$$

where ε is the 2-dimensional anti-symmetric tensor to make the condensate a Lorentz-scalar operator, and $I_{ab,ij} = \delta_{ab} J_{ij}$, because $r \otimes \bar{r}$ contains a unique singlet. In principle J_{ij} could be anything unless we bring in further constraints from the strong interaction dynamics. There are supporting arguments [*51*] based on the idea of large N_c expansions that J_{ij} must be such that the unbroken group H_f is maximal (we shall briefly outline this argument in Appendix A). Thus $J_{ij} = \delta_{ij}$ in which case clearly $H_f = SU(N) \otimes U(1)$, for $SU(N)$ group preserves a sesquilinear metric, which can be reduced to the unit matrix.

Case 2: ψ transforms according to a real IRR under G_S. Thus there exist N fermion multiplets

$$\psi_\alpha{}^{(r)ai} \in r(G_S). \tag{2.14}$$

Therefore the chiral flavor group is $G_f = SU(N)$. The general form of the condensate is

$$\Sigma = \langle \varepsilon_{\alpha\beta} \psi_\alpha{}^{(r)ai} \psi_\beta{}^{(r)bj} I_{ab,ij} \rangle \neq 0. \tag{2.15}$$

Now $I_{ab,ij} = S_{ab} J_{ij}$ with $S_{ab} = S_{ba}$ because the symmetric product of any two real representations contains a singlet. Again the maximum flavor preservation is obtained for $J_{ij} = \delta_{ij}$ yielding for condensate

$$\Sigma = \langle \varepsilon_{\alpha\beta} \psi_\alpha{}^{(r)ai} \psi_\beta{}^{(r)bi} S_{ab} \rangle \neq 0 \tag{2.16}$$

which breaks the symmetry to $SO(N)$ ($SO(N)$ is the group which preserves a bilinear symmetric metric which can be reduced to a unit matrix).

Case 3: ψ transforms according to pseudo-real IRR of G_S. Since any pseudo-real representation is of even dimension, say $2N$, then the chiral group is $G_f = SU(2N)$. In this case the G_S-singlet is contained in the anti-symmetric part of the cross product of a pseudo-real representation with itself, hence $I_{ab,ij} = A_{ab} J_{ij}$, where $A_{ab} = -A_{ba}$. Again the maximum flavor symmetry preservation is achieved for

$$J_{ij} = \left(\begin{array}{c|c} & I \\ \hline -I & \end{array} \right)_{2N \times 2N} \tag{2.17}$$

(because $A_{ab} = -A_{ba}$ necessarily implies $J_{ij} = -J_{ij}$) This J definies a bilinear antisymmetric metric preserved by the symplectic group. Thus $H = \mathrm{Sp}\,(2N)$.

Case 4: This is the general case, which is an arbitrary combination of the first three cases. Each piece in this case condenses separately (perhaps not all at once, but one after another in the same order of mass scale. We will discuss this case more thoroughly in the last section of this review, devoted to the idea of Tumbling).

Here let us emphasize again that the statements obtained above crucially depend on the assumption that the condensate must be a bilinear one (instead of quadrilinear for instance) and the symmetry breaking preserves the maximum possible flavor subgroup (see Appendix A).

PESKIN [*12*] made a very elegant observation that in each of the 3 basic cases discussed above that the broken part of the chiral flavor group G_f/H_f is a symmetric space [*14*]. If we denote the unbroken generators by T_i, and the broken generators by X^a,

$$\begin{aligned} H_f &\leftrightarrow \{T_i\} \\ G_f/H_f &\leftrightarrow \{X_a\} \end{aligned} \tag{2.18}$$

this means that there exist a parity symmetry operation which preserves the Lie algebra of G_f such that

$$\begin{aligned} PT_iP^{-1} &= T_i \\ PX_aP^{-1} &= -X_a. \end{aligned} \tag{2.19}$$

Of course another equivalent criterion for the symmetric space is that of X and T represent arbitrary linear combinations of broken and unbroken generators, their commutators satisfy

$$\begin{aligned} [T, T] &= iT \\ [X, T] &= iX \\ [X, X] &= iT. \end{aligned} \tag{2.20}$$

We shall normalize these generators so that

$$\begin{aligned} \mathrm{Tr}\,(T_iT_j) &= \delta_{ij} \\ \mathrm{Tr}\,(X_aX_b) &= \delta_{ab} \\ \mathrm{Tr}\,(X_aT_i) &= 0. \end{aligned} \tag{2.21}$$

A useful representation for the generators of G_f in the general case is:

$$\begin{aligned} T_i &= \left(\begin{array}{c|c} A & B \\ \hline B & -A^T \end{array}\right), \qquad A = A^+ \\ X_a &= \left(\begin{array}{c|c} C & D \\ \hline D^+ & C^T \end{array}\right), \qquad C^+ = C,\ \mathrm{Tr}\, C = 0. \end{aligned} \tag{2.22}$$

For the three special cases discussed above we have

$$\begin{aligned} &\text{Case 1:} \quad B = 0, \quad D = 0 \\ &\text{Case 2:} \quad B = -B^T, D = D^T \\ &\text{Case 3:} \quad B = B^T, \quad D = -D^T. \end{aligned} \tag{2.23}$$

Now we are equipped to work out the Vacuum Alignment problem explicitly:

We fix the embedding $H \subset G$, by choosing a specific representation for the broken and unbroken generators X_a and T_i. We choose $|0\rangle$ as a reference vacuum from the set $\{|0, g\rangle\}$. Next we turn on the electroweak interactions. Denoting those generators of G_f coupled to the gauge fields of G_w by ϑ we describe this perturbation by

$$\Delta H = -\sum_{\vartheta} A_\mu{}^\vartheta J_\mu{}^\vartheta \tag{2.24}$$

where $J_\mu{}^\vartheta = \bar{\psi}\gamma_\mu\vartheta\psi$, and ψ is purely left handed. ϑ are linear combinations of the generators of G_f, and are defined to contain the coupling constants:

$$\vartheta = \{\vartheta_T, \vartheta_x \mid \vartheta_T \in \mathscr{H}, \vartheta_x \in \mathscr{G}\}. \tag{2.25}$$

Since we are going to search for the correct vacuum by varying the orientation of G_w in G_f, the relevant perturbation is

$$\Delta H(U) = \boldsymbol{U}^+(g)\, \Delta H \boldsymbol{U}(g) = -\sum A_\mu{}^{U^+\vartheta U} J_\mu{}^{U^+\vartheta U}. \tag{2.26}$$

Thus the energy shift for $|0\rangle$ due to the exchange of weak gauge bosons, for different orientations of $G_w(g) \subset G_f$ are

$$\Delta\varepsilon(U) = -\frac{1}{2} \sum \int d^4x\, D_{\mu\nu}(x)\, T\langle 0|\, J^{U^+\vartheta U}(x)\, J^{U^+\vartheta U}(0)\, |0\rangle \tag{2.27}$$

where $D_{\mu\nu}(x)$ is the massless vector boson propagator.
Now we shall make use of the assumption that G/H is a symmetric space. There is a very important consequence of this property which we are going to state as a theorem.

Theorem. The only H-invariant term in the product $X_a X_b$ is proportional to δ_{ab} except in the trivial case that G/H is a direct product of symmetric spaces.
Proof follows immediately from the observation that a set of Hermitian matrices X_a cannot be decomposed into two distinct representations of H, except in the trivial case.
It follows from this theorem, and (2.21), that for the correct vacuum, $|0\rangle$, the two-current correlation functions satisfy:

$$\begin{aligned} T\langle 0|\, J_\mu{}^{X_a} J_\nu{}^{X_b}\, |0\rangle &= \delta^{ab} T\langle 0|\, J_\mu{}^X J_\nu{}^X\, |0\rangle = \mathrm{Tr}\,(X^a X^b)\, T\langle 0|\, J_\mu{}^X J_\nu{}^X\, |0\rangle \\ T\langle 0|\, J_\mu{}^{T_i} J_\nu{}^{T_j}\, |0\rangle &= \delta_{ij} T\langle 0|\, J_\mu{}^T J_\nu{}^T\, |0\rangle = \mathrm{Tr}(X_i X_j)\, T\langle 0|\, J_\mu{}^T J_\nu{}^T\, |0\rangle. \end{aligned} \tag{2.28}$$

Where in these equations the notation X and T denote any single generator X_a and T_i (no sum over these generators). Using the definition relation (2.19) we also obtain:

$$T\langle 0|\, J_\mu{}^{T_i} J_\nu{}^{X^a}\, |0\rangle = 0. \tag{2.29}$$

We may develop the corresponding relations for the two-current correlation function of the electro-weak currents. Using $\mathrm{Tr}\,(XT) = 0$, and (2.25), we can write

$$\begin{aligned} T\langle 0|\, J_\mu{}^{U^+\vartheta U} J_\nu{}^{U^+\vartheta U}\, |0\rangle = {} & \mathrm{Tr}\,[(U^+\vartheta U)_T\, (U^+\vartheta U)_T]\, T\langle 0|\, J_\mu{}^T J_\nu{}^T\, |0\rangle \\ & + \mathrm{Tr}\,[(U^+\vartheta U)_X\, (U^+\vartheta U)_X]\, T\langle 0|\, J_\mu{}^X J_\nu{}^X\, |0\rangle. \end{aligned} \tag{2.30}$$

Now we can simplify this relation by using

$$\vartheta_T \vartheta_T = \vartheta^2 - (\vartheta_T \vartheta_X + \vartheta_X \vartheta_T) - \vartheta_X \vartheta_X \tag{2.31}$$

$$\begin{aligned} \mathrm{Tr}\,[(U^+\vartheta U)_T\, (U^+\vartheta U)_T] &= \mathrm{Tr}\,[(U^+\vartheta U)\, (U^+\vartheta U)] - \mathrm{Tr}\,[(U^+\vartheta U)_X\, (U^+\vartheta U)_X] \\ &= \mathrm{Tr}\,[\vartheta^2] - \mathrm{Tr}\,[(U^+\vartheta U)_X\, (U^+\vartheta U)_X]. \end{aligned} \tag{2.32}$$

Thus, two terms combine to give

$$T\langle 0|\, J_\mu^{U^+\vartheta U} U J_\nu^{U^+\vartheta U}\, |0\rangle = \mathrm{Tr}\,[\vartheta^2]\, T\langle 0|\, J_\mu^T J_\nu^T\, |0\rangle + \mathrm{Tr}\,[(U^+\vartheta U)_X\, (U^+\vartheta U)_X]\, T\langle 0|\, J_\mu^X J_\nu^X - J_\mu^T J_\nu^T\, |0\rangle \quad (2.33)$$

and

$$\Delta\varepsilon(U) = -\frac{1}{2} \sum_\vartheta \int d^4x D_{\mu\nu}(x)\, T\langle 0|\, J_\mu^T(x)\, J_\nu^T(0)\, |0\rangle + \frac{1}{2} \sum_\vartheta \mathrm{Tr}\,[(U^+\vartheta U)_X\, (U^+\vartheta U)_X] \int d^4x D_{\mu\nu}(x)\, T\langle 0|\, J_\mu^T J_\nu^T - J_\mu^X J_\nu^X\, |0\rangle. \quad (2.34)$$

Preskill [13] has extensively used spectral-function sumrules to show that the quantity in bracket is positive. Thus the preferred vacuum is obtained by minimizing the quantity

$$F[U] = \frac{1}{2} \sum_\vartheta \mathrm{Tr}\,[(U^+\vartheta U)_X\, (U^+\vartheta U)_X] \quad (2.35)$$

over possible equivalent representations of X_a and T_i.
There is a striking by-product which follows this criterion, which we state as a theorem:

Theorem: The preferred orientation of G_w relative to H is that which breaks the gauge symmetry the least; it is the orientation which minimizes the vector boson masses.

Proof: We recall that a typical signal for SSB is the creation of NGB from the vacuum (true vacuum) by a broken generator. The relation $\langle 0|\, T(J^T J^X)\, |0\rangle = 0$ guarantees the equality of the corresponding NGB decay constants and then

$$\langle 0|\, J_\mu^{X_a}(0)\, |\Pi^b(p)\rangle = ip_\mu F_\pi \delta^{ab} = ip_\mu F_\pi\, \mathrm{Tr}\,(X^a X^b). \quad (2.36)$$

For dynamically broken theories the vacuum polarization tensor at zero momentum behaves as

$$\Pi_{\mu\nu}^{\vartheta'\vartheta}(q) \propto T\langle 0|\, J_\mu^{\vartheta'}(q)\, J_\nu^{\vartheta}(-q)\, |0\rangle \underset{q\to 0}{\simeq} \left(g_{\mu\nu} - \frac{q_\mu q_\nu}{q^2}\right) M_{\vartheta'\vartheta}^2. \quad (2.37)$$

Comparing these two equations and saturating the two-current correlation function by the corresponding NGB's, we get

$$M_{\vartheta'\vartheta}^2 = \sum_a F_\pi^2\, \mathrm{Tr}\,(X^a\vartheta')\, \mathrm{Tr}\,(x^a\vartheta) = F_\pi^2\, \mathrm{Tr}\,(\vartheta_X{}'\vartheta_X). \quad (2.38)$$

In the last step we have used the completeness of $\{X_a\}$. Using the fact that a general vacuum $|0, g\rangle$ and the NGB states built upon it are related to the corresponding states of $|0\rangle$ by a unitary transformation, we find that for the preferred vacuum $|0, g\rangle$,

$$M_{\vartheta\vartheta'}^2 = F_\pi^2\, \mathrm{Tr}\,[(U^+\vartheta U)_X\, (U^+\vartheta' U)_X]. \quad (2.39)$$

Thus we finally see that the energy shift generated by the electro-weak perturbation is proportional to the trace of the dynamically generated mass matrix of the electro-weak gauge bosons. That completes the proof.
We can state the important result of this theorem in several different ways: a) The embedding $H \subset G$ can be characterized by a fermion condensate which forms in the channel in which the lowest order G_w interaction is the most attractive, b) The preferred orien-

tation of G_w relative to H is such that the largest possible subgroup of G_w survives (gauge symmetries resist being broken).
After this important detour let us proceed to identify the state $|0\rangle$ which minimizes $\Delta\varepsilon(U)$, and state the following simple theorem.

Theorem. Any given orientation of G_w relative to H is at least a stationary point of $\Delta\varepsilon$ if $[\vartheta_T, \vartheta_X] = 0$.

Proof. Writing $U = \exp(iw^a X_a)$, and noting

$$\frac{\partial}{\partial U} F[U] = \frac{\partial}{\partial w^a} F[U(w)]\Big|_{\substack{w^a=0\\(U=I)}} . \tag{2.40}$$

We get after some trivial algebra

$$\frac{\partial}{\partial w^a} F[U]\Big|_{U=I} = -i \operatorname{Tr}([X^a, \vartheta]_X \vartheta_X) = -i \operatorname{Tr}(X^a[\vartheta_T, \vartheta_X]) = 0. \tag{2.41}$$

The stability of this vacuum requires a study of the curvature of $\Delta\varepsilon$ about it. Let us assume that this particular one is the preferred vacuum (if not it is going to yield tachyonic NGB's). We have observed above that every direction in G/H away from the true vacuum along which $\Delta\varepsilon$ has a non-zero curvature corresponds to a massive particle. The mass matrix of these massive excitations of NGB's was obtained above as

$$M^2_{ab} = \frac{1}{F_a F_b} \frac{\partial^2}{\partial w^2\, \partial w^b} \Delta\varepsilon(U)\Big|_{w=0} .$$

Writing $\Delta\varepsilon(U) = F[U]\, I$, we obtain

$$\frac{\partial^2}{\partial w^a\, \partial w^b} F[U]\Big|_{w^a=0} = -\operatorname{Tr}([X_a, \vartheta)_X [X_b, \vartheta]_X) - \operatorname{Tr}\left([X_a, [X_b, \vartheta]]_X \vartheta_X\right). \tag{2.42}$$

The first term can be simplified using

$$[X_a, \vartheta]_X = [X_a, \vartheta_T + \vartheta_X]_X = [X_a, \vartheta_T]_X + [X_a, \vartheta_X]_X .$$

The only surviving term in $[X_a, \vartheta]_X [X_b, \vartheta]_X$ is

$$[X_a, \vartheta_T]_X [X_b, \vartheta_T]_X = [X_a, \vartheta_T] [X_b, \vartheta_T] .$$

All the other three terms involve $[X_a, \vartheta_X]_X$ which vanishes identically (because $[X_a, \vartheta_X]$ is a T-like generator). Using trace property we can rewrite the first term as

$$\operatorname{Tr}([X_a, \vartheta_T] [X_b, \vartheta_T]) = -\operatorname{Tr}\left([\vartheta_T, [\vartheta_T, X_a]] X_b\right). \tag{2.43}$$

The second term involves the triple commutator

$$[X_a, [X_b, \vartheta]]_X = [X_a, [X_b, \vartheta_T + \vartheta_X]]_X = [X_a, [X_b, \vartheta_T]]_X + [X_a, [X_b, \vartheta_X]]_X . \tag{2.44}$$

Since $[X_b, \vartheta_T]$ is a X-like generator, the first triple commutator is T-like, hence it vanishes. Since the second triple commutator is purely X-like, we get

$$[X_a, [X_b, \vartheta]]_X = [X_a, [X_b, \vartheta_X]] . \tag{2.45}$$

We once again use the trace property to rewrite the second term as

$$\mathrm{Tr}\left([X_a, [X_b, \vartheta]]_X \vartheta_X\right) = \mathrm{Tr}\left([X_a, [X_b, \vartheta_X]] \vartheta_X\right) = \mathrm{Tr}\left([\vartheta_X, [\vartheta_X, X_a]] X_b\right). \quad (2.46)$$

Combining these we obtain,

$$\frac{\partial^2}{\partial w^a \partial w^b} F[U]\Big|_{w^a=0} = \sum \mathrm{Tr}\left([\vartheta_T, [\vartheta_T, X_a]] X_b - [\vartheta_X, [\vartheta_X, X_a]] X_b\right) \quad (2.47)$$

and thus finally

$$M_{ab}^2 = \frac{M^2}{4\pi} \sum \mathrm{Tr}\left\{[\vartheta_T, [\vartheta_T, X_a]] X_b - [\vartheta_X, [\vartheta_X, X_a]] X_b\right\} \quad (2.48)$$

where

$$M = \frac{4\pi}{F_\pi{}^2} I = \frac{4\pi}{F_\pi{}^2} \int d^4x D_{\mu\nu}(X) \, T\langle 0| \, J_\mu{}^T(x) \, J_\nu{}^T(0) - J_\mu{}^X(x) \, J_\nu{}^X(0) \, |0\rangle. \quad (2.49)$$

There is one important lesson which comes out from this discussion: While checking to see which one is the true vacuum, we automatically obtained the masses of the PNGB's. A tachyonic PNGB is a sign of vacuum instability.
Since we need to have $M^2 > 0$ for a stable vacuum, let us discuss the sign of each factor in (2.49) carefully.

1. The matrices $D^{ab} = [\vartheta, [\vartheta, X^a]] X^b$ are positive matrices (have positive eigenvalues). Diagonal elements can be written as squares of the structure constants f^{abc}. Thus unbroken generators θ_T give a contribution to the mass matrix M_{ab}^2 which is strictly positive, whereas broken generators give negative contributions which tend to destabilize the vacuum.
2. Above we used the information that $M^2 > 0$. Preskill [13] gave a detailed discussion of this point using arguments based on Weinberg's spectral function sum rules. We shall briefly review his arguments in Appendix B.

Next we shall illustrate these formal arguments on a simple toy model.

c) A Simple Example

We shall shortly see that any realistic model which should give masses to quarks and leptons, along with $W^\mp$ and Z, must have enlarged group structure to include weak coupling between technifermions (which have acquired dynamical masses through breaking of their chiral symmetries) and the ordinary fermions. The existence of at least 3 doublets each of quarks and leptons will invariably lead to at least 2 doublets of technifermions. To give a first impression of how complex these realistic models could get, let us briefly discuss the vacuum alignment problem for a Technicolor model involving two doublets of technifermions, postponing the promised weak coupling to the ordinary quarks ad leptons to the next section:

$$Q_1 = \begin{pmatrix} U \\ D \end{pmatrix}, \qquad Q_2 = \begin{pmatrix} C \\ S \end{pmatrix}. \quad (2.50)$$

In order to avoid extra $SU(2)$ symmetries linking these doublets we need to distinguish them. In the standard theory this is done by assigning doublets different Yukawa couplings to fundamental scalars. Here we don't have any fundamental scalars. Thus we distinguish them by assigning them electric changes differing by an amount δ. Now denoting the average charges of the doublets by q_1, and q_2, the hypercharge assignments become:

$$Q_{1L} = \begin{pmatrix} U \\ D \end{pmatrix}_L, \quad Q_{2L} = \begin{pmatrix} C \\ S \end{pmatrix}_L, \quad U_R, \quad D_R, \quad C_R, \quad S_R \tag{2.51}$$

$$\frac{Y}{2}: \quad q_1 \quad q_2 \quad q_1 + \frac{1}{2}, \quad q_1 - \frac{1}{2}, \quad q_2 + \frac{1}{2}, \quad q_2 - \frac{1}{2}.$$

Now defining a 4-component multiplet as

$$Q = \begin{pmatrix} Q_1 \\ Q_2 \end{pmatrix}_{4\times 1} \tag{2.52}$$

the A_μ-interaction is easily obtained as

$$\mathcal{L}_{(A)} = g A_\mu{}^a \bar{Q}_L \gamma_\mu (\tau^a \otimes 1)\, Q_L . \tag{2.53}$$

The notation will be explained below.
Finding the form of B_μ-interaction requires a little work:

$$\begin{aligned} \mathcal{L}_{(B)} &= g' B_\mu [2q_1 \bar{Q}_{1L} \gamma_\mu Q_{1L} + 2q_2 \bar{Q}_{2L} \gamma_\mu Q_{2L} \\ &\quad + (2q_1 + 1)\, \bar{U}_R \gamma_\mu U_R + (2q_1 - 1)\, \bar{D}_R \gamma_\mu D_R \\ &\quad + (2q_2 + 1)\, \bar{C}_R \gamma_\mu C_R + (2q_2 - 1)\, \bar{S}_R \gamma_\mu S_R] \\ &= g' B_\mu [2q_1 \bar{Q}_{1L} \gamma_\mu Q_{1L} + 2q_2 \bar{Q}_{2L} \gamma_\mu Q_{2L} \\ &\quad + 2q_1 \bar{Q}_{1R}\, \gamma_\mu Q_{1R} + 2q_2 \bar{Q}_{2R}\, \gamma_\mu Q_{2R} \\ &\quad + \bar{Q}_{1R} \tau^3 \gamma_\mu Q_{1R} + \bar{Q}_{2R} \tau^3 \gamma_\mu Q_{2R}] . \end{aligned} \tag{2.54}$$

We can simplify this, by noting that;

$$\begin{aligned} & 2q_1 \bar{Q}_{1L} \gamma_\mu Q_{1L} + 2q_2 Q_{2L} \gamma_\mu Q_{2L} \\ &= (q_1 + q_2)\, (\bar{Q}_{1L} \gamma_\mu Q_{1L} + \bar{Q}_{2L} \gamma_\mu Q_{2L}) + (q_1 - q_2)\, (\bar{Q}_{1L} \gamma_\mu Q_{1L} - \bar{Q}_{2L} \gamma_\mu Q_{2L}) \\ &= (q_1 + q_2)\, \bar{Q}_L \gamma_\mu (1 \otimes 1)\, Q_L + (q_1 - q_2)\, \bar{Q}_L \gamma_\mu (1 \otimes \tau^3)\, Q_L \end{aligned} \tag{2.55}$$

and similarly,

$$\begin{aligned} & 2q_1 \bar{Q}_{1R} \gamma_\mu Q_{1R} + 2q_2 \bar{Q}_{2R} \gamma_\mu Q_{2R} \\ &= (q_1 + q_2)\, \bar{Q}_R \gamma_\mu (1 \otimes 1)\, Q_R + (q_1 - q_2)\, \bar{Q}_R \gamma_\mu (1 \otimes \tau^3)\, Q_R \end{aligned} \tag{2.56}$$

$$\begin{aligned} \mathcal{L}_{(B)} &= g' B_\mu \{(q_1 + q_2)\, \bar{Q} \gamma_\mu (1 \otimes 1)\, Q + \bar{Q}_R \gamma_\mu (\tau_3 \otimes 1)\, Q_R \\ &\quad + (q_1 - q_2)\, [\bar{Q}_L \gamma_\mu (1 \otimes \tau_3)\, Q_L + \bar{Q}_R \gamma_\mu (1 \otimes \tau_3)\, Q_R]\} . \end{aligned} \tag{2.57}$$

Now let us write everything in terms of two component Weyl spinors:

$$(Q_R)^c = \begin{pmatrix} (Q_{1R})^C \\ (Q_{2R})^C \end{pmatrix} \tag{2.58}$$

and

$$\begin{aligned} \mathcal{L}_{(B)} &= g' B_\mu \{(q_1 + q_2)\, [\bar{Q}_L \gamma_\mu (1 \otimes 1)\, Q_L - (\overline{Q_R})^c\, \gamma_\mu (1 \otimes 1)\, (Q_R)^c] \\ &\quad - (\overline{Q_R})^c\, \gamma_\mu (\tau_3 \otimes 1)\, (Q_R)^c \\ &\quad + \delta [Q_L \gamma_\mu (1 \otimes \tau_3)\, Q_L - (\overline{Q_R})^c\, \gamma_\mu (1 \otimes \tau_3)\, (Q_R)^c]\} . \end{aligned} \tag{2.59}$$

Now we can express the electro-weak perturbation in terms of the totally left-handed Weyl-spinor language:

$$\psi = \begin{pmatrix} Q_L \\ (Q_R)^C \end{pmatrix} = \begin{pmatrix} Q_{1L} \\ Q_{2L} \\ (Q_{1R})^C \\ (Q_{2R})^C \end{pmatrix} \tag{2.60}$$

$$\Delta\mathcal{L}_{(w)} = A_\mu{}^a J_\mu{}^a + B_\mu J_\mu \tag{2.61}$$

where

$$J_\mu{}^a = \bar{\psi}\gamma_\mu \vartheta^a \psi; \qquad J_\mu = \bar{\psi}\gamma_\mu \vartheta \psi$$

$$\frac{1}{g}\,\vartheta^a = \left(\begin{array}{c|c} \tau_a \otimes 1 & \\ \hline & 0 \end{array}\right) \tag{2.62}$$

$$\frac{1}{g'}\,\vartheta = (q_1 + q_2)\left(\begin{array}{c|c} 1 \otimes 1 & \\ \hline & -1 \otimes 1 \end{array}\right) = \left(\begin{array}{c|c} 0 & \\ \hline & -\tau_3 \otimes 1 \end{array}\right) + \delta\left(\begin{array}{c|c} 1 \otimes \tau_3 & \\ \hline & -1 \otimes \tau_3 \end{array}\right).$$

In the limit where the electro-weak perturbation is turned off, the chiral flavor group is $G_f = SU(4)_L \otimes SU(4)_R \otimes U(1)$. In the purely left-handed formalism the generators of G_f are 8×8 matrices which divide into 4×4 blocks. Each block acts on a vector $Q = (U\,C\,D\,S)$ and thus is expressed as a direct product of 2×2 weak isospin matrix and a 2×2 matrix in the space of two doublets. This was the motivation behind the form of the final expression for the electroweak perturbation given in (2.61).

We assume that $G_f = SU(4)_L \otimes SU(4)_R \otimes U(1)$ breaks down to $H_f = SU(4)_V \otimes U(1)$. In the two-component notation the embedding $SU(4)_V \subset SU(4)_L \otimes SU(4)_R$ is characterized by the fermion condensate which can be represented by a 8×8 matrix of the form (where U is the same representation of $g \in G_f$ which we have employed to change the orientation of $G_w \subset G_f$ in the first version):

$$\Sigma = \left(\begin{array}{c|c} 0 & U \\ \hline U^T & 0 \end{array}\right)_{8 \times 8}. \tag{2.63}$$

In this specific example we choose to follow the orthogonal approach to the one presented in the general discussion of the vacuum alignment where we had fixed the embedding $H_f \subset G_f$, and searched for the stationary vacuum by varying the orientation of $G_w \subset G_f$. Here we assume that the embedding $G_w \subset G_f$ is fixed, that is the generator of G_w given in (2.62) correspond to the true vacuum. (In this equivalent alternative version) the group theoretical factor in the effective potential becomes

$$F[U] = -\frac{1}{2}\,\mathrm{Tr}\,(\Sigma^+ \vartheta \Sigma \vartheta^T) \tag{2.64}$$

where

$$\vartheta = \begin{pmatrix} \lambda_L & \\ & -\lambda_R{}^* \end{pmatrix} \tag{2.65}$$

and λ_L, λ_R are the generators of the G_w in the 4-component notation. This form allows as a further interpretation of the so-called principle of "Resistance to Breaking": The alignment $H_f \subset G_f$ is determined by the requirement that the value of G_w Casimir operator $\vartheta\vartheta^T$ acting on the condensate Σ is as small as possible. In particular, if it is possible for the condensate to be oriented so that it is G_w-invariant this orientation always corresponds to a minimum of the effective potential; the G_w group doesn't break unless it must:

The minimum of the effective potential can be shown to occur at the following stationary points

$$U_1 = \left(\begin{array}{cc|cc} 1 & 0 & & \\ 0 & 1 & & \\ \hline & & 1 & 0 \\ & & 0 & 1 \end{array}\right) \qquad \Sigma_1 = \langle \bar{U}_R U_L + \bar{D}_R D_L + \bar{C}_R C_L + \bar{S}_R S_L \rangle \tag{2.66}$$

$$U_2 = \left(\begin{array}{cc|cc} 1 & 0 & 0 & 0 \\ 0 & 0 & 1 & 0 \\ \hline 0 & 1 & 0 & 0 \\ 0 & 0 & 0 & 0 \end{array}\right) \qquad \Sigma_2 = \langle \bar{U}_R U_L + \bar{C}_R C_L + \bar{D}_R C_L + \bar{S}_R S_L \rangle . \tag{2.67}$$

Before computing the explicit form of the effective potential, let us first observe that the A-interactions do not contribute to $\Delta\varepsilon$ to the lowest order. We can intuitively understand this as follows: The NGB's couple to the axial currents, but we can use right-handed currents as their interpolating currents. The $SU(2)_w$ currents commute with these interpolating currents. (Thus the $SU(2)_w$ interaction cannot give mass to NGB's.) Σ_1 leaves $U(1)_{e.m}$ unbroken, but Σ_2 breaks G_w completely. The effective potential at these stationary points can be worked out explicitly now:

$$\begin{aligned} \Delta\varepsilon[\Sigma_1] &= \frac{g'^2}{4\pi} M^2(2q_1^2 + 2q_2^2) \\ \Delta\varepsilon[\Sigma_2] &= \frac{g'^2}{4\pi} M^2[(q_1 + q_2)^2 + q_1 - q_2] . \end{aligned} \tag{2.68}$$

Computing the difference, we get

$$\Delta\varepsilon[\Sigma_1] - \Delta\varepsilon[\Sigma_2] = \frac{g'^2}{4\pi} M^2[(q_1 - q_2)(q_1 - q_2 - 1)] = \frac{g'^2}{4\pi} M^2\delta(\delta - 1) . \tag{2.69}$$

Thus if $0 < \delta < 1$, Σ_2 minimizes the effective potential, and G_w is completely broken. This result is quite striking: Although in this model there exists a relative orientation of G_w and H_f which leaves $U(1)_{e.m}$ unbroken, this doesn't correspond to the true vacuum if $|\delta| < 1$. This means that two technidoublets of slightly different charge will not line up and leave the photon massless. This clearly shows that, in building Technicolor models the charge assignments for the technifermions, are extremely crucial.

Now we shall compute the PNGB spectrum. First we divide the ϑ into unbroken and broken parts. With the convention described in (2.22),

$$\vartheta_T{}^a = \frac{g}{2}\left(\begin{array}{c|c} \tau_a \otimes 1 & \\ \hline & -\tau_a{}^T \otimes 1 \end{array}\right); \qquad \vartheta_X{}^a = \left(-\frac{g}{2}\right)\left(\begin{array}{c|c} \tau_a \otimes 1 & \\ \hline & \tau_a{}^T \otimes 1 \end{array}\right) \tag{2.70}$$

and

$$\begin{aligned} \vartheta_T &= g'(q_1 + q_2)\left(\begin{array}{c|c} 1 \otimes 1 & \\ \hline & -1 \otimes 1 \end{array}\right) + \frac{g'}{2}\left(\begin{array}{c|c} \tau_3 \otimes 1 & \\ \hline & -\tau_3 \otimes 1 \end{array}\right) + g'\delta\left(\begin{array}{c|c} 1 \otimes \tau_3 & \\ \hline & -1 \otimes \tau_3 \end{array}\right) \\ \vartheta_X &= \frac{g'}{2}\left(\begin{array}{c|c} \tau_3 \otimes 1 & \\ \hline & -\tau_3 \otimes 1 \end{array}\right) . \end{aligned} \tag{2.71}$$

The 15 NGB's of the $SU(4)_L \otimes SU(4)_R/SU(4)_V$ correspond to the generators

$$X^a = \begin{pmatrix} C & \\ & C^T \end{pmatrix}, \quad C^+ = C, \ \mathrm{Tr}\, C = 0. \tag{2.72}$$

In the basis where the condensate is Σ_1, let us divide them into two classes:

Class 1 (7 NGB):

$$C = \tau^a \otimes 1 \div \{\bar{U}\gamma_5 D + \bar{C}\gamma_5 S, \bar{D}\gamma_5 U + \bar{S}\gamma_5 C, (\bar{U}\gamma_5 U - \bar{D}\gamma_5 D) + (\bar{C}\gamma_5 C - \bar{S}\gamma_5 S)\}$$

$$C = \tau^a \otimes \tau_3: \{\bar{U}\gamma_5 D - \bar{C}\gamma_5 S, \bar{D}\gamma_5 U - \bar{C}\gamma_5 S, (\bar{U}\gamma_5 U - \bar{D}\gamma_5 D) - (\bar{C}\gamma_5 C - \bar{S}\gamma_5 S) \tag{2.73}$$

$$C = 1 \otimes \tau_3: (\bar{U}\gamma_5 U + \bar{D}\gamma_5 D) - (\bar{C}\gamma_5 C + S\gamma_5 S).$$

Class 2 (8 NGB):

$$C = \tau^a \otimes \tau^{1,2}, \quad \begin{cases} \bar{U}\gamma_5 C, \bar{D}\gamma_5 S, \bar{U}\gamma_5 S, \bar{D}\gamma_5 C \\ \bar{C}\gamma_5 U, \bar{S}\gamma_5 D, \bar{S}\gamma_5 U, \bar{C}\gamma_5 D. \end{cases} \tag{2.74}$$

As was remarked above also, while using our mass formula for the NGB's we come across with a striking feature:

1. A_μ-interactions don't give masses to any of these 15 NGB's. Because the two terms in (2.48) cancel exactly.
2. For the NGB's in class 1, again there is a pairwise cancellation between the broken and unbroken generators of $U(1)$. Thus these NGB's don't receive a mass to lowest order.
3. Only the NGB's of the class 2 receive a mass via B_μ-exchange:

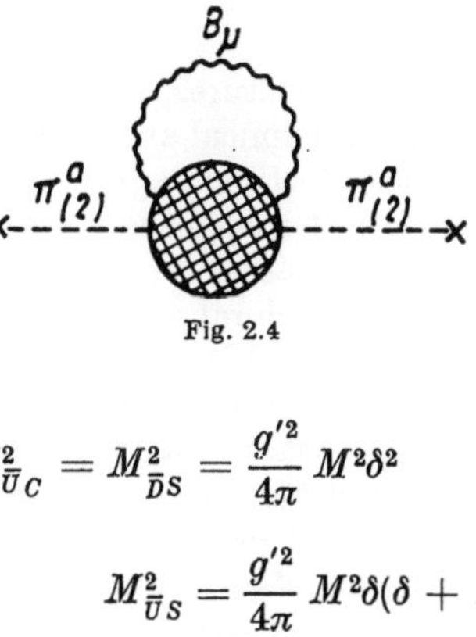

Fig. 2.4

$$M^2_{\bar{U}C} = M^2_{\bar{D}S} = \frac{g'^2}{4\pi} M^2\delta^2$$

$$M^2_{\bar{U}S} = \frac{g'^2}{4\pi} M^2\delta(\delta + 1) \tag{2.75}$$

$$M^2_{\bar{D}C} = \frac{g^{2\prime}}{4\pi} M^2\delta(\delta - 1).$$

Once again we see that if $|\delta| < 1$, the vacuum we have chosen (Σ_1) is not stable, although this choice leaves photon massless.

Had we chosen to work with the other condensate, then G_w would be totally broken; thus it does not make any sense to use this model any further if $|\delta| < 1$.

The physics we have discussed so far has included fermions and gauge bosons without fundamental scalars, transforming under the gauge group $U(1)_y \otimes SU(2)_L \otimes SU(3)_c \otimes G_{TC}$. This, no doubt, cannot be a complete description in particular it fails to account for masses of light fermions. Since light fermions' mass term is not chiral invariant (for them to acquire mass) the chiral invariance of light fermions must be broken. Evidently new interactions may be required. These new interactions should also contribute to the masses of the PNGB's and should eliminate any experimentally inacceptable massless NGB's (acions). The new interactions are postulated to occur at a variety of scales all about the TC-scale. In the next section we will show how this can give rise to a complex pattern of fermion masses below the TC-scale.

3. Extended Technicolor

a) Introduction

By introducing the technifermions, which transform under a new gauge group G_{TC} which becomes strong at about 1 TeV, we have successfully accounted for the breaking of the electro-weak group $G_w \equiv SU(2)_w \otimes U(1)_Y$. This procedure was a scaled up version of the dynamical chiral symmetry breaking by the QCD condensates, and it produced the masses of the gauge bosons $W^\pm$, and Z, leaving the photon exactly massless. It was further possible to guarantee the relation $M_Z = M_W \cos\theta_W$ to all orders in the strong G_{TC} and the color interactions, by a global dynamically unbroken $SU(2)$ group, which we called the custodial $SU(2)$.

Despite all this success of the TC theories, there is one very striking failure of them, namely the DSB mechanism does not give ordinary fermions any mass at all. The ordinary fermions and technifermions have separate chiral symmetris, and the part of this symmetry belonging to the ordinary fermions remains unbroken by the TC condensates, thus preventing them having masses. We know that the technifermions gain dynamical masses of order $\mu_{TC} \approx 1$ TeV, but the ordinary fermions gain no mass at all, dynamical or current.

Due to the elementary nature of these lectures, we would like to digress for a moment and clearly state what we mean by dynamical and current masses. The current mass of a fermion is the mass term in some (effective) Lagrangian in front of the term $\bar{\psi}\psi$. Since this mass appears in the Lagrangian, it is also the mass that appears in PCAC type relations, upon which the name current mass was given. The dynamical mass [*15*], on the other hand is a momentum dependent chiral symmetry breaking term in the fermion propagator which behaves like μ^3/q^2 for $q^2 \gg \mu^2$, for asymptotically-free theories. This dynamical mass is an artifact of the strong bound state effects and it disappears a $q^2 \to \infty$. For example the dynamical mass of all three light quarks is about 300 MeV, roughly a third of the nucleon mass. A third term often used is constituent mass, which is the sum of the dynamical and current masses. The constituent masses of the quarks are roughly equal to one half of the $q\bar{q}$ states formed out of them[2]).

It is then clear that we have to generate current masses for the quarks, and a large global chiral symmetry is preventing us from doing so. Let's demonstrate this explicitly: We have the following fermions

$$\begin{pmatrix} u \\ d \end{pmatrix}_L^\alpha \begin{pmatrix} c \\ s \end{pmatrix}_L^\alpha \begin{pmatrix} \nu_e \\ e \end{pmatrix}_L \begin{pmatrix} \nu_\mu \\ \mu \end{pmatrix}_L, \dots$$

$$u_R^\alpha,\ c_R^\alpha,\ \dots$$

$$d_R^\alpha,\ s_R^\alpha,\ e_R,\ \mu_R,\ \dots$$

[2]) Using the current masses $2\text{ MeV} \leqq M_u < M_d \leqq 15\text{ MeV}$, $m_s \simeq 160$ MeV, $M_c \simeq 1.3$ GeV, $M_b \sim 4.6$ GeV, and a dynamical mass of about 300 MeV for all of them, we see that the constituent masses of the quarks are roughly one half of the masses of ϱ, Φ, Ψ, Γ.

possibly more[3]). The index α above is the color index, which we can take to run over 1, 2, and 3.[4]). The global chiral symmetry group of these fermions, when the color and the electro-weak (EW) interactions are urned off, contains the group $SU(16)_L \otimes SU(14)_R \otimes U(1)$. When turned on, the color interactions explicitly break this chiral group, and the unbroken subgroup contains $(SU(4)_L \otimes SU(4)_R)_q \otimes (SU(4)_L \otimes SU(2)_R)_l \otimes U(1)$ where the indices q and l refer to quarks and leptons. This symmetry is further broken by $SU(2)_W$ explicitly to a subgroup containing $(SU(2)_L \otimes SU(4)_R)_q \otimes (SU(2)_L \otimes SU(2)_R)_l \otimes U(1)$. And finally, $U(1)_Y$ explicitly breaks it to a subgroup containing $(SU(2)_L \otimes SU(2)_R \otimes SU(2)_R)_q \otimes (SU(2)_L \otimes SU(2)_R)_l \otimes U(1)$. Clearly this extra global symmetry is unbroken explicitly by any interaction as yet introduced. Furthermore, it cannot be dynamically broken either; otherwise this will produce TNGB which will remain exactly massles to all orders. Therefore this symmetry will prevent the leptons and quarks from acquiring masses.

Until now two possible solutions have been presented for this problem. In these lectures we will only present the first [*17, 18, 19*] and not the most recent one [*20*]. Both of these solutions require that these chiral symmetries be broken explicitly. In the first scheme (historically the older one), a new gauge interaction, called extended technicolor (ETC) [*17, 18*], or sideways group [*19*], is introduced which explicitly breaks these extra chiral symmetries. The second scheme is more economical, and it uses the electro-weak interactions to break this symmetry through instanton effects. We will not discuss this second scheme any further in these lectures, and we will discuss the first one in a model independent way.

b) Model Independent Analysis of ETC Theories

It is clear that the intention for introducing the ETC group is to gauge the flavor symmetry of the fermions, more precisely to gauge the electro-weak generations. Precisely for that reason it is very natural to take $[SU(2)_W, G_{ETC}] = 0$, i.e., the W and ETC interactions are orthogonal to each other. Even though this scheme is the only natural extention of the electro-weak structure of the known light fermions, it is by no means a requirement on the theory [*19*]. If we choose to put the right handed technifermions in EW doublets and the left handed ones in singlets, then, $[SU(2)_W, G_{ETC}) \neq 0$, if the ETC interactions are to connect fermions and technifermions, the necessity of which we shall prove shortly. The models in which $[SU(2)_W, G_{ETC}] \neq 0$, however, are clearly not as aesthetically appealing as those where EW and ETC interactions are orthogonal. In the latter alternative something unnatural has been inserted in these models, where the quarks and techniquarks behave differently under the electro-weak group. In the rest of these lectures we'll limit ourselves to the case $[SU(2)_W, G_{ETC}] = 0$.

Let us now consider the mechanism, by which the ETC interactions will give current masses to the fermions, as shown in figure 3.1. This one loop contribution to the current masses of the fermions appears to be logarithmically divergent as usual, but actually it converges due to the nature of the dynamical mass. We will remind the reader again, at this point, that for $q^2 \gg \mu^2$, the dynamical mass term goes to zero like μ^3/q^2, and the

---

[3]) We know at least two others, namely τ and b, but the electro-weak partners of these, namely ν_τ and t have not yet been observed experimentally. Also the question of, whether or not the right handed components of ν_e and ν_μ exist, has been revived by recent experiments and a number of theoretical papers [*16*].

[4]) We take $G_c = SU(3)$, and all the quarks to be triplets. This fact certainly is not known beyond the s quark; but it seems to be the only straightforward generalization. For example a model in which quark transforms as a color sextet will look very unnatural, will require additional fermions or unusual charge assignments to avoid triangle anomalies, and it will be extremely difficult to fit them in any grand unified scheme.

chiral mixing disappears at high q^2. In other words, the momentum dependence of the dynamical mass acts like a cut off and renders the momentum integral finite. The result of this integration is then

$$M_q \sim \left(\frac{g_{ETC}^2}{8\pi^2}\right) \frac{\mu_{TC}^3}{M_{ETC}^2}\, \Gamma, \tag{3.1}$$

where $\mu_{TC} \sim 1$ TeV, the scale at which TC condensates form, and the scale of the dynamical mass of the fermions, M_{ETC} is the mass of the ETC gauge bosons, and Γ is some constant (including all the mixing angles) of order one. We see immediately that[5])

$$M_{ETC} \sim 20\ \text{TeV} - 40\ \text{TeV}$$

will yield about 1 GeV, which we consider to be typical lepton or quark mass that we want to produce.

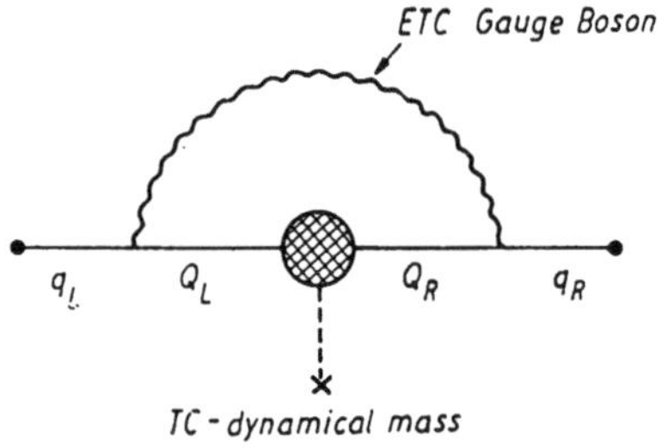

Fig. 3.1

We mentioned before, that we have to couple quarks and techniquarks together through the ETC gauge bosons. As a matter of fact, we implicitly assumed this in the above calculation. Now let's show that this is really so, and there is no alternative. To see that consider the diagram in figure 3.2:

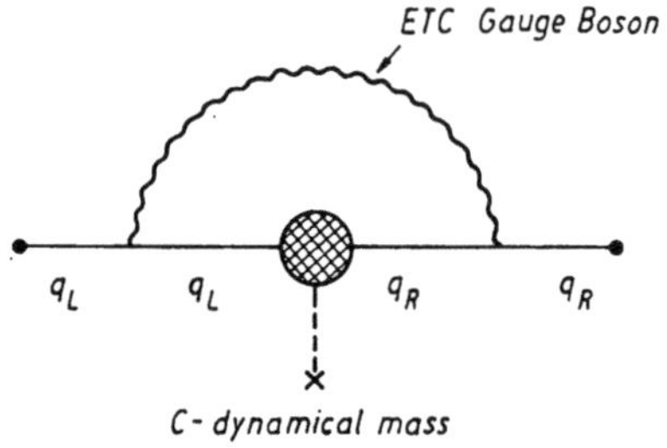

Fig. 3.2

where we replaced Q by q, and the dynamical mass μ_{TC} by μ_C. Then we obtain again

$$M_q \sim \left(\frac{g_{ETC}^2}{8\pi^2}\right) \frac{\mu_C{}^3}{M_{ETC}^2}\, \Gamma .$$

Substituting $\mu_C \sim 0.5$ GeV, $M_q \sim 1$ GeV, we obtain $M_{ETC} \sim 0.3$ GeV, which is *phenomenologically unacceptable.*

---

[5]) We assume that at the mass scale μ_{TC}, $(g_{ETC}^2/4\pi) < 1$. When we show that $G_{TC} \subset G_{ETC}$, this relation will become automatic since at μ_{TC}, $(g_{TC}^2/4\pi) \sim 1$.

In short then, *we conclude that the ETC gauge bosons, must couple ordinary fermions to technifermions, and they must have both left and right handed couplings.* We assume also, that there are no elementary scalars in the theory at any level; therefore, the masses of the ETC gauge bosons are generated dynamically by another interaction that we name technicolor-prime (TC'), which becomes strong at a mass scale several times the ETC gauge boson mass, typically say at 100 TeV or thereabouts.
This result has an immediate corrollary: since the fermions are TC singlets and the technifermions are non-singlets, to connect these two together ETC interactions must not be orthogonal to TC interactions, i.e., $[G_{ETC}, G_{TC}] \neq 0$. This result in turn implies that either G_{TC} is a subgroup of $G_{ETC}(G_{TC} \subset G_{ETC})$, or else there is a group $\overline{G}$ which is dynamically broken by TC', and such that, $G_{TC} \subset \overline{G}$, $G_{ETC} \subset \overline{G}$, and $[G_{ETC}, G_{TC}] \neq 0$, within $\overline{G}$[6]). TC' interactions dynamically break the group $\overline{G}$ to H, where $G_{TC} \subset H$. In this case we can call $\overline{G}$ the ETC group by a change of terminology, since it is broken by the TC' interactions.

In conclusion we see that either $G_{TC} \subset G_{ETC}$, or else G_{ETC} must be enlarged so that $G_{TC} \subset G_{ETC}$. From this point on, then, we will use, without loss of generality $G_{TC} \subset G_{ETC}$.

Next we would like to prove a very important result first observed by EICHTEN and LANE [*19*], concerning the reducibility of the fermion representations. Before stating it as a theorem and then producing a proof, let's first introduce some notation that will be useful in the proof: Let G' be the group $G_{ETC} \otimes U(1)_X$ if $G_C \equiv SU(3) \subset G_{ETC}$, if not let it be the group such that $G' = \overline{G} \otimes U(1)_X$, where $\overline{G} \supset G_{ETC}$ and $\overline{G} \supset G_C$[7]), and $U(1)_X$ is some $U(1)$ group and in general $U(1)_Y$ piece of G_{EW} is a linear combination of $U(1)_X$ and some of the generators of $\overline{G}$. Let D_L be the representation of the left handed fermions under the full local group $SU(2)_W \otimes G'$. Let D_R be the corresponding representation of all right-handed fermions. We know that D_R is reducible under $SU(2)_W$[8]), therefore it has at least two irreducible components under $SU(2)_W \otimes G'$, the full gauge group. Now we are ready to state and prove the following theorem:

Theorem: D_L must be irreducible, and D_R must contain exactly two irreducible components.

Proof: Let us write

$$D_L = \sum_{j=1}^{r_L} \oplus D_L^{(j)}$$

$$D_R = \sum_{j=1}^{r_R} \oplus D_R^{(j)}$$

where $D_L^{(j)}$ and $D_R^{(j)}$ are irreducible representations of $SU(2)_W \otimes G'$ of dimensions $n_L^{(j)}$ and $n_R^{(j)}$ respectively. Clearly then

$$\dim(D_L) \equiv N_L = \sum_{j=1}^{r_L} n_L^{(j)}$$

$$\dim(D_R) \equiv N_R = \sum_{j=1}^{r_R} n_R^{(j)}.$$

---

6) $G_{ETC} \subset G_{TC}$ is ruled out, since a broken group cannot be the subgroup of an unbroken group.
7) We must warn the reader we are using the notation $\overline{G}$ in a different way than we did when we proved above $G_{TC} \subset G_{ETC}$, this is a different $\overline{G}$.
8) We assumed that the right handed fermions and technifermions were $SU(2)_W$ singlets. As was mentioned above, for the case $[SU(2)_W, G_{ETC}] \neq 0$, such is not true.

All of these fermions must be TC' singlets, otherwise they would gain dynamical masses of order 100 TeV[9]). Since the singlet is a real representation of any group, the chiral symmetry of these fermions is $SU(N_L + N_R)$ before any of the other interactions are turned on. When we turn on $\bar{G}$ interactions (remember $\bar{G} \supseteqq G_{ETC}$, $\bar{G} \supset G_C$) this global chiral symmetry is explicitly broken to a subgroup $G^{(1)} \subset SU(N_L + N_R)$, such that

$$G^{(1)} \supset \underbrace{\big(SU(2) \otimes \cdots \otimes SU(2)\big)_L}_{r_L \text{ terms}} \otimes \underbrace{\big(SU(2) \otimes \cdots \otimes SU(2)\big)_R}_{r_R \text{ terms}} \otimes U(1)$$

$G^{(1)}$ is actually equal to this subgroup if no irreducible component $D_L^{(j)}$ or $D_R^{(j)}$ occurs more than once in D_L and D_R. Next we turn the $SU(2)_W$ interactions on, and this break $G^{(1)}$ explicitly to $G^{(2)}$, where

$$G^{(2)} \supset \underbrace{\big(U(1) \otimes \cdots \otimes U(1)\big)_L}_{r_L \text{ terms}} \otimes \underbrace{\big(SU(2) \otimes \cdots \otimes SU(2)\big)_R}_{r_R \text{ terms}} \otimes U(1).$$

And finally, turning $U(1)_X$ on breaks $G^{(2)}$ explicitly to $G^{(3)}$, where

$$G^{(3)} \supset \left(\prod_{j=1}^{r_L} \otimes U(1)\right)_L \otimes \left(\prod_{j=1}^{r_R} \otimes U(1)\right)_R \otimes U(1).$$

On the other hand, if want to give mass to at least one member of all the electro-weak doublets, and require (as demanded by phenomenology) that these masses be different, then the maximum global explicitly and dynamically unbroken symmetry we can tolerate is $U(1) \otimes U(1) \otimes U(1) \otimes U(1)$[10]) corresponding to linear combinations of baryon, lepton, technibaryon and technilepton number[11]). Furthermore, we cannot break any explicitly unbroken symmetry dynamically for this will produce TNGB, which remain massless to all orders in the gauged interactions. Therefore, the only way out of this is to have

$$r_L + r_R + 1 \leqq 4.$$

If we combine this with the requirement $r_L \geqq 1$, $r_R \geqq 2$, we conclude that

$$r_L = 1, \quad \text{and} \quad r_R = 2,$$

thus proving the theorem.
It is worth mentioning that a slightly weaker version of this theorem holds even if $[SU(2)_W, G_{ETC}] \neq 0$. In that case we can easily conclude again

$$r_L + r_R + 1 \leqq 4, \qquad r_L \geqq 1, \qquad r_R \geqq 1,$$

which has three possible solutions: $(r_L = 1, r_R = 1)$, $(r_L = 1, r_R = 2)$, $(r_L = 2, r_R = 1)$. Now let's go back to the first version of this theorem with $[SU(2)_W, G_{ETC}] = 0$, and state some important and obvious corollaries:

First of all, since all the left handed fermions must be put in a single irreducible representation, in particular the quarks and leptons must be in this representation together, thus *implying quark-lepton unification at an energy scale of about 100 TeV*. Of

[9]) 100 TeV is a buzzword for a mass scale which is several times that of the masses of the *ETC* gauge bosons, or the constant $F_{\pi'}$ of the pions of the TC' fermions.
[10]) Some of these global $U(1)$ symmetries may suffer from the triangle anomaly.
[11]) In general all of these are violated individually, but certain linear combinations of them remain unbroken.

course, this implies baryon number violation by the ETC interactions; however in most models it is possible to implement the exact stability of the proton. This is done as follows: Let N_l and N_b be the lepton and baryon numbers. The model has four (up to anomalies) exactly conserved global charges, N_1, N_2, N_3, and N_4. If it is such that,

$$N_1 + N_b = \text{(linear combination of others, but not of } N_l\text{)},$$

$$N_2 + N_l = \text{(linear combination of others, but not of } N_b\text{)},$$

then even though N_b and N_l are both violated by ETC interactions, all the physical processes below the energy threshold of technifermions ($\sim$1 TeV) conserve both lepton and baryon number separately, sine no technifermion can appear in the final state. *In particular the ETC gauge boson exchange makes no contribution to the proton decay.* Secondly note that $\overline{G} = G_{ETC}$ if $G_C \subset G_{ETC}$. If G_C is not contained in G_{ETC} then $\overline{G}$ is some group that contains both G_C and G_{ETC} and must be broken by TC' interactions. Therefore repeating the argument we gave for ETC, we can say that we have to change our notation and call G the ETC group. *In conclusion we say that either $G_C \subset G_{ETC}$ or else G_{ETC} must be enlarged so that $G_C \subset G_{ETC}$.* From now one we will use, without loss of generality, that $G_C \subset G_{ETC}$. This is actually a stronger result than that of Eichten and Lane [*19*], who observed that $[G_{ETC}, G_C] \neq 0$. We have to make one point clear now. We saw that either $G_C \otimes G_{TC} \subset G_{ETC}$, or else, it is necessary to enlarge G_{ETC} such that this is so. However, such an enlargement costs us something: It is no longer necessary for G_{ETC} to break to $G_C \otimes G_{TC}$ in one step. The breaking can proceed in different ways:

$$\text{(a)}\ G_{ETC} \xrightarrow{100\,\text{TeV}} G_C \otimes G_{TC},$$

$$\text{(b)}\ G_{ETC} \xrightarrow{100\,\text{TeV}} G_{TC} \otimes H \xrightarrow{30\,\text{TeV}} G_{TC} \otimes G_C,$$

$$\text{(c)}\ G_{ETC} \xrightarrow{100\,\text{TeV}} G_C \otimes H \xrightarrow{30\,\text{TeV}} G_{TC} \otimes G_C.$$

The diagrams of various running coupling constants of these schemes are shown in figures 3.3, 3.4, and 3.5.

A third comment can be made in regard to the groups $U(1)_{e.m.}$, $U(1)_Y$, and $U(1)_X$. First note that since the leptons carry integer charges and quarks carry fractional charges, and leptons and quarks are put in the same multiplet of $G_{ETC} \otimes SU(2)_W \otimes U(1)_X$[12]), and since there are quarks and leptons with both $T_3 = 1/2$ and $T_3 = -1/2$ under $SU(2)_W$, we conclude that $[G_{ETC}, U(1)_{e.m.}) \neq 0$. The generator of $U(1)_{e.m.}$ namely the electric charge can be written as $\hat{Q} = \hat{F}_3 + \hat{Y}$, where $\hat{F}_3$ is the diagonal generator of $SU(2)_W$, and $\hat{Y}$ is the weak hypercharge, i.e., the generator of $U(1)_Y$. Since leptons and quarks

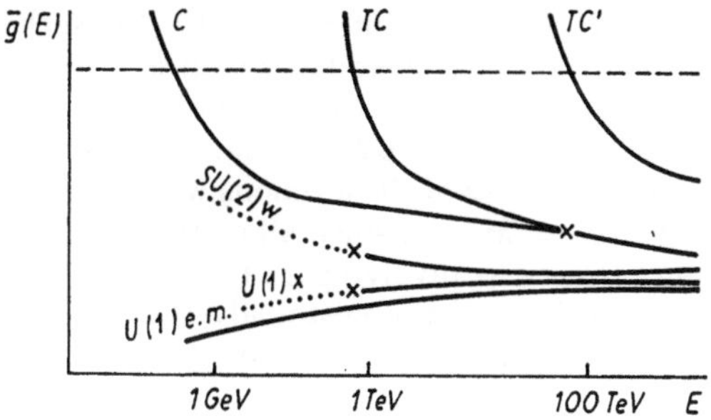

Fig. 3.3

12) From now on we drop the notation $\overline{G}$ altogether following the argument above.

have different weak hypercharges, $\hat{Y}$ must be a linear combination of $\hat{X}$ the generator of $U(1)_X$ and the diagonal generators of G_{ETC}. The relative orientations of G_{ETC}, $SU(2)_W$, $U(1)_X$, $U(1)_Y$, $U(1)_{e.m.}$ are shown in figure 3.6.

Next we'd like to address ourselves to the question of fermion masses in a little more detail. The question we have in mind concerns the custodial $SU(2)$ group, which was defined to be some global dynamically unbroken subgroup of the chiral symmetry, under

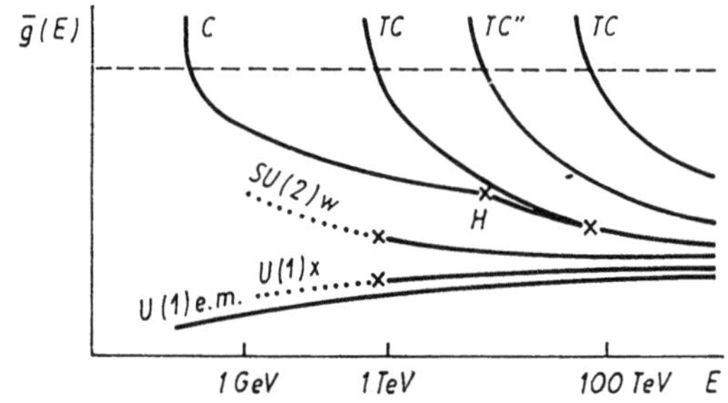

Fig. 3.4

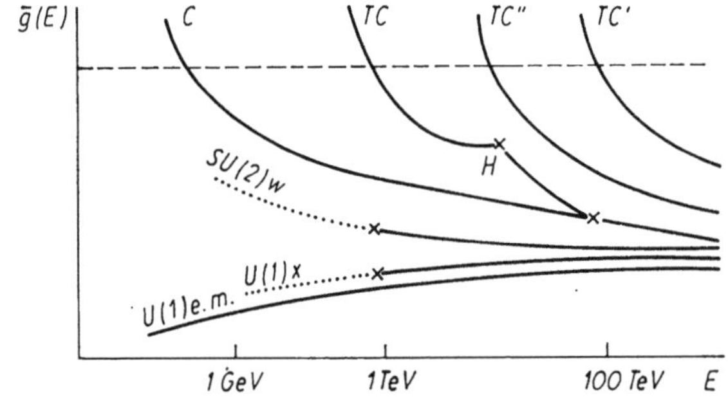

Fig. 3.5

which the generators of $SU(2)_W$ transform as a triplet. The existence of such an $SU(2)$ group guarantees us that the relation

$$M_W = M_Z \cos \theta_W$$

is exact to all orders in TC and color interactions[13]). Since the final surviving global symmetry is a product of $U(1)s$, the custodial $SU(2)$ is broken (partly or completely) explicitly by EW and ETC interactions. As a matter of fact we can see that the custodial $SU(2)$ has to be explicitly broken by the ETC interactions for a different reason. Without such an explicit breaking the masses of the fermions within the same EW doublet will be identical, i.e., $M_u = M_d$, $M_c = M_s$, ..., since all of these masses are produced by one loop (and higher order) corrections involving ETC gauge boson exchange. In this case we see that the explicit breaking of the custodial $SU(2)$ allows us to have $M_u \neq M_d$, without introducing large corrections to the relation $M_W = M_Z \cos \theta_W$. Including all the

[13]) We emphasize that the custodial $SU(2)$ must be unbroken not only by TC condensates, but also color condensates as well. In the relation $M_W = M_Z \cos \theta$, we can tolerate "percentish" corrections (such as EW), but not strong (color) corrections.

corrections then, we can write it as

$$M_W = M_Z \cos\theta_W \left(1 + O(\alpha) + O\left(\alpha \sum_i \frac{m_+^{(i)^2} - m_-^{(i)^2}}{M_W^2}\right)\right) \tag{3.2}$$

where i labels various flavors (including leptons) and (+) and (−) signs refer to the $T_3 = 1/2$ and the $T_3 = -1/2$ weak isospin members respectively. Assuming that there are no badly split (the same oder of M_W) doublets, all of these corrections are within a percent or two and in agreement with the experimental value of 0.985 ± 0.023 [*3*]. Another point about the *ETC* theories which can deal with in a model independent way concerns the *TC'* interaction. One wonders about how detailed a knowledge of this sector is needed in order to describe the low energy region. Reassuringly, the answer is "very

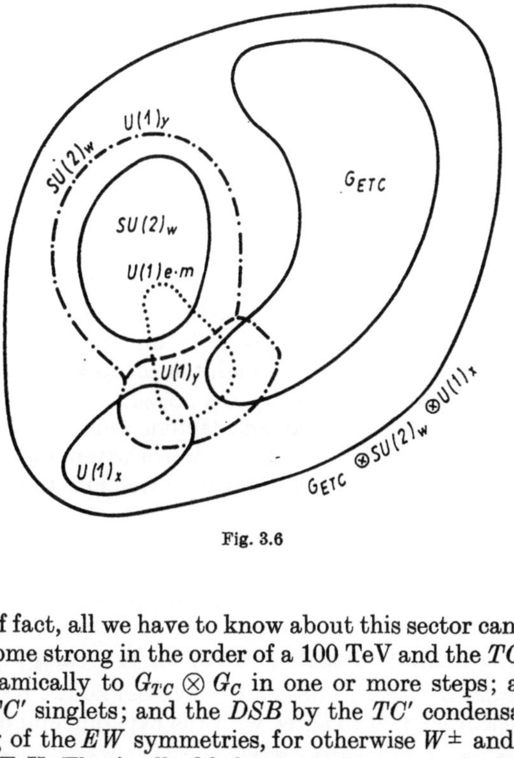

Fig. 3.6

little". As a matter of fact, all we have to know about this sector can be listed as follows: *TC'* interactions become strong in the order of a 100 TeV and the *TC'* condensates break the *ETC* group dynamically to $G_{TC} \otimes G_C$ in one or more steps; all the fermions and technifermions are *TC'* singlets; and the *DSB* by the *TC'* condensates does not contribute to the breaking of the *EW* symmetries, for otherwise $W^\pm$ and Z masses will be of the order of tens of TeV. That's all: Of these requirements, the last one has an interesting corollary, which says that the *TC'*-fermions must have no weak interactions at all, or vectorial (parity conserving) weak interactions [*19*]. It is obvious that the *DSB* by *TC'* condensates cannot contribute to the breaking of the *EW* group if the *TC'*-fermions do not have weak interactions. The other case however, is not so trivial. To demonstrate that they are allowed to have vector *EW* interactions, take the doublets (one flavor only)

$$\begin{pmatrix} U' \\ D' \end{pmatrix}_L^\alpha \begin{pmatrix} U' \\ C' \end{pmatrix}_R^\alpha, \qquad \alpha = 1, \ldots, N,$$

where α is the *TC'* index. The global symmetry is as usual $SU(2)_L \otimes SU(2)_R \otimes U(1)$ if the *TC'* representation is complex, or $SU(4)$ if it is real. This global symmetry is broken dynamically[14]) to $SU(2)_{L+R} \otimes U(1)$, or $SO(4)$ or $Sp(4)$ depending on the TC' group an the reality of its representations, where the index $L + R$ represents the diagonal (vectorial) $SU(2)$. At this point, it is further necessary to note that $SO(4) = SO(3) \otimes SO(3) \cong SU(2) \otimes SU(2)$, and $Sp(4) \cong SO(5) \supset SO(4) \cong SU(2) \otimes SU(2)$. The global flavor symmetry will be explicitly broken by the EW interactions completely, only the local group $SU(2)_V \otimes U(1)$ remaining, where V denotes "vectorial".

As shown above, in all three scenarios, the dynamically unbroken subgroup of the flavor symmetry includes the explicitly unbroken (local) group. (Actually for the first (SU) scenario of DSB, the dynamically and explicitly unbroken groups are identical.) Therefore, there are no generators of the EW group which are dynamically broken by the TC' condensates. This fact was first observed by EICHTEN and LANE [19] for the SU scenario of DSB. The proof above, can of course be repeated for the case when there are more than one flavors of the TC' fermions.

Before we leave the topic of the TC' interactions altogether, we would like to point out that the reader does not have to worry about the PNGB of the symmetries dynamically broken by TC' condensates and explicitly broken by $G_{ETC} \otimes SU(2)_w \otimes U(1)$. Typically these PNGB will be 100 times heavier than those which are associated with the dynamically broken symmetries by the TC condensates, the lightest ones having masses of order no less than 500 GeV.

To conclude our model independent analysis of the ETC theories let us now summarize the phenomenological restrictionson such theories first, and the direct implications of these on the acceptable models afterwards.

These phenomenological restrictions (most of which are obvious) are as follows:

a) The light particles (TC and TC' singlets) must include at least three EW generations of quarks and leptons, and the gauge bosons γ, $W^{\pm}$ and Z.

b) The relation $M_W = M_Z \cos\theta_W$ must be maintained up to all orders for the interactions of all non-abelian unbroken groups $G_C, G_{TC}, G_{TC'}, \ldots$

c) Reasonable current masses for quarks and leptons must be obtained, in particular the mass splittings within an EW doublet ($M_u \neq M_d$) or within the successive generations ($M_u \neq M_c$) must be generated without introducing corrections to the relation $M_W = M_Z \cos\theta_W$ by more than a few percent.

d) Proton must be stable; more precisely it's life time must be longer than the present experimental lower bound of about $\approx 10^{30}$ years[15]).

e) There has to be no axion (either a TNGB or a very light PNGB) of which mass is less than 2 GeV [23].

f) The flavor changing neutral currents (FCNC) for low energy phenomena must be suppressed. In particular G_{ETC} contribution to processes like K^0, $\bar{K}^0$ mixing, $K_L \to \mu^+\mu^-$, etc. ... must be sufficiently small to agree with the experimental data.

Other than these strictly phenomenological restrictions, we add four more, which are theoretical, but which are just as essential for a successful ETC model:

g) There has to be no local ABJ [24] triangle anomalies at any stage of the theory.

h) The theory must not suffer from the strong CP violation [25].

i) The fermion multiplets must not be too large to destroy asymptotic freedom.

j) The preferred vacuum state (vacuum alignment) of the flavor and gauge groups must not yield disastrous results, such as a massive photon.

k) No elementary scalars should appear at any stage, at least up to about 10^{15} GeV, where presumably, the grand unification takes place.

---

[14]) Of course we don't know how it is broken. But if we assume, as explained in part II, the maximum flavor preservation, then that's how the DSB proceeds.

[15]) For the present experimental status see the review articles of References [21] and [22].

Well, that is at all order! It is not surprising that no one has yet come up with a model that does all of that. There are semi-realistic or toy models [*17, 18, 26*] that exhibit some of these properties. In particular the problem of strong *CP* violation has not been solved to everybody's satisfaction yet [*27–30*]. There are problems with the FCNC [*29*]. The role of the global anomalies is not fully understood. The instanton contributions (if important at all) are not known. However the remedies for the other restrictions above have been found, as described in this section. (The problem in model building now, is to find a model that satisfies all of these conditions at once, it is possible to find models which satisfy these conditions individually, or some of them at a time).
Now, let's list then, the results of our model independent analysis, namely the consequences of these restriction on the *ETC* theories:

1. To avoid TNGB's, no global subgroup of the flavor group of fermions must be dynamically broken if it is explicitly unbroken.
2. Dynamically broken subgroup of the flavor group by the color condensates must be global (otherwise, we'll have gauge bosons which are as light as 100 MeV).
3. All the chiral symmetry of the model must be explicitly broken down to the products of $U(1)$ groups. In particular the unbroken global subgroup (dynamically and explicitly) can not be larger than $U(1) \otimes U(1) \otimes U(1) \otimes U(1)$ (up to global $U(1)$ anomalies).
4. There has to be a global $SU(2)$ subgroup of the flavor group, which is dynamically unbroken, but which is explicitly broken by the *ETC* and *EW* interactions. We call this subgroup the custodial $SU(2)$, and it guarantees that the corrections to $M_W = M_Z \cos \theta_W$ are no more than a few percent, while allowing large fermion splittings ($M_u \neq M_d$).
5. All left handed fermions must form an irreducible representation of G_{ETC}. All right handed fermions must be put in no more than two irreducible representations of G_{ETC}.
6. If the right handed counterparts of the $T_3 = 1/2$ and $T_3 = -1/2$ left handed light fermions belong to two separate right handed irreducible representations of G_{ETC}, then these two representations cannot be identical, for otherwise all the Cabibbo type mixing angles will vanish.

For reasons that will become clear in the next part, we will content ourselves here with this model independent analysis of the *ETC* models, and we won't take time to study any specific models.

4. Tumbling

a) Introduction

As appealing as they are for having given us a glimpse of a world without elementary scalars, the *TC* and *ETC* theories have their drawbacks. Here, we don't intend to dwell upon those points, which (in principle) can be solved and understood (some day, when we acquire a better understanding of strong interactions). But rather, it is our intention to point out some aesthetically unappealing features, which make the theory bulgy and cumbersome. This is a bit like what happened with the Higgs mechanism, where the scalars had to proliferate too rapidly in any reasonable realistic grand unified theory (GUT). The *ETC* models are severely restricted as we saw in the previous section; but they are open to proliferation in some other respect, namely new strong interactions of TC, TC', TC'' ... are introduced at higher and higher mass scales and they are loosely connected with each other. For instance, the reader should remember how little was enough to know about the TC' group for model building in a simple minded *ETC* theory.

Furthermore, such a proliferation of these sequential strong groups is almost a requirement, otherwise there would be no natural explanation for the drastically different masses of quark and lepton generations ranging from 1 MeV[16]) to perhaps 50 GeV[17]). One might be content to pass up this point by remembering that the theories with elementary scalars have nothing intelligent to say about this point either. In these theories the fermion mass hierarchy is generated by a hierarchy of Yukawa couplings that range from 10^{-6} to 10^{-1} for the ordinary fermions, without even the remotest understanding of why that should be so. However one of the strongest motivations for starting the development of the dynamically broken theories was the desire to have a theory with no unnatural numbers or adjustments in it. The *ETC* theories can solve this problem, but the cost is high: One has to introduce many sequential strong groups.
Next, we'll take a different aspect of the theory, namely the need for unification. In this day and age, it is hardly necessary for us to explain in detail why a theory based on a group with many simple groups occurring in it as cross products cannot be expected to be a fundamental theory. It cries for unification! However, how do we unify a dynamically broken theory? If we stick to the rules of the game we introduced so far, such a unification is not possible: Each time to break a (larger) symmetry one needs a new strong group, which itself must remain unbroken. It is a never-ending chain, unless another mechanism takes over at some mass scale, such as Higgs mechanism or perhaps another dynamical mechanism. Perhaps there are Higgs scalars, after all, above the grand unification mass (10^{15} GeV), and the symmetry way up there is broken by them, and the surviving symmetries are later dynamically broken at lower mass scales. We find this explanation quite unsatisfying, though perhaps not wrong. Isn't there an alternative where grand unification is still possible, completely within the context of dynamical symmetry breaking, with no need for scalars, ever? If such an alternative is possible after all, it would imply that there must be a dynamical mechanism by which a gauge group can break itself, not just some piece of the global chiral symmetry of the fermions in it. The tumbling hypothesis first introduced by Raby, Susskind, and Dimopoulous [*31*] is an attempt to provide an answer to this need. This hypothesis, though very appealing and plausible, is beyond the reach of our ability to verify by calculation, due to the insufficiency of our knowledge of how to compute with strong interactions.

b) Tumbling Hypotheses

To lead to the hypotheses of tumbling, let's consider an asymptotically free gauge theory with fermions. At high enough of a mass scale M, the coupling constant must be weak owing to the asymptotic freedom. With the decreasing mass scale, the running coupling constant $g(\mu)$ increases. When it becomes strong enough, the fermion bound states form in the scalar (as well as in other) channels. The masses of these bound states decrease with increasing coupling constant. At some point the coupling constant becomes so strong that some of these scalar bound states becomes massless. At this point then, Raby, Susskind and Dimopoulous hypothesized that there may be a critical value of the running coupling constant $g_c = g(\mu_c)$ attained at the mass scale μ_c, beyond which a massless scalar multiplet(s) condensates. The multiplet, which is favoured to condensate before any others is clearly the most tightly bound one, and this line of argument led Raby, Susskind, and Dimopoulous to introduce the following first hypothesis of tumbling:

---

16) That of course gets worse if the neutrinos are massive with $m_\nu \leqq 100$ eV.
17) The present experimental search has failed to detect the top quark up to 37 GeV of center of mass energy in e^+e^- collisions.

Hypothesis 1

In an asymptotically free gauge theory with an anomaly free fermion content[18]*, there is a critical value g_c of the running coupling constant attained at the mass scale μ_c. At a scale below μ_c the most strongly bound scalar (pseudo scalar) multiplet of the bound states condensates thus breaking the symmetry spontaneously. If this condensate is not a singlet under G, the group itself is broken. The gauge bosons of the broken generators and the fermions which participate in the condensate gain masses of order μ_c.*

Let's elaborate on this assumption a little. First of all, we must make it clear immediately that as before, we are using the two component Weyl spinors here. The usage of the Weyl spinors is not a mere convenience but a necessity, which will become clear when we explain another fact, namely the fermion representation content must be non-real. So, now let's put all fermions (and antifermions) in two component Weyl spinors. Let the representation be denoted by D, where D may possibly be reducible. Since we are dealing with an asymptotically free gauge theory, at low enough of a mass scale, the coupling constant will become strong enough to form scalar bound states. The simplest of these bound state scalars will be di-fermion states. Possibly the bound states of 4 or 6, or in general $2n$ fermions may form. With decreasing mass scale, however, the coupling constant will continue to increase, since the theory is so far unbroken. According to this hypothesis, then, there exists a critical value of the coupling constant, and at mass scales below the critical mass scale at which the coupling constant attains its critical value, the most strongly bound scalar multiplet will condensate. Again the simplest possibility is a di-fermion condensate. In this case one of the irreducible components of $D \otimes D$ will be the most strongly bound multiplet and it will condensate. This multiplet may be a singlet under the gauge group G, in which case the group G will remain unbroken, and some of the global chiral symmetries will be broken. This is the usual dynamical chiral symmetry breaking mechanism we started with; this is what happens with the color group, and presumably what happens with technicolor as well. On the other hand if the most strongly bound multiplet in $D \otimes D$ is not a singlet under G, then clearly G itself will be broken by the condensate. This is the reason for the non-reality of the fermion representation. This result, (that the product representation $D \otimes D$ contains no singlets[19]) is a necessary condition but not sufficient.

Here we would like to make two comments about the need for a non-real representation D, so that G may break itself. The first of these comments is about a sufficiency condition so that $D \otimes D$ should not contain any singlets at all. The answer is easy enough if we first note the fact that the non-reality of D is both necessary and sufficient if D is irreducible. Therefore a necessary and sufficient condition in general is that any subset of the irreducible components of D must be non-real. The second comment concerns the multi-fermion condensates. For those, this condition is no longer sufficient (though still necessary). For example if $G = SU(4)$ and if D is the fundamental (4) representation of $SU(4)$ (which is non-real), then the four fermion condensate $qqqq$ which transforms as $4 \otimes 4 \otimes 4 \otimes 4$ of $SU(4)$ has a singlet in it.

[18]) All representations of the groups $SU(2)$, $SO(2n+1)$, $SO(4n)$, $Sp(2n)$, G_2, F_4, E_7 and E_8 are either real or pseudo-real, thus anomaly free. $SO(4n-2)$ for $n \geqq 2$ [*32*] and E_6 [*33*] are also shown to have no triangle anomaly either, even though they have some complex representations. Therefore the triangle anomaly is restricted to the groups $U(1)$ and $SU(n)$ for $n \geqq 3$. However, all the self conjugate (such as the adjoint) representation of $SU(n)$ are anomaly free. Various complex but reducible representations of $SU(n)$ are also free from the ABJ anomaly and the examples of that will be given in this section and the following one.

[19]) To the contrary if D is real, $D \otimes D$ has at least one singlet in it.

This first tumbling hypothesis is clearly very appealing and intuitively easy to accept. But nevertheless it is still an assumption, for we do not know, if such a critical value of the coupling constant exists for each gauge group and for each fermion representation. We must emphasize here, that such a critical value of the coupling constant g, will be a function of the fermion representation content. Let us say, when the fermions transform with D the critical value of g is g_c and is attained at a mass scale μ_c. Instead, if the fermions transformed by a different representation D', we'd know that $g(\mu_c)$ will no longer be equal to the critical value g_c. However, we want to emphasize that, the naive guess that $g(\mu'_c) = g_c$ will determine the new mass scale at which the condensate occurs need not be true at all. In general we should write $g(\mu'_c) = g'_c$ and $g_c \neq g'_c$. Clearly assuming the existence of such a critical value for all groups and all fermion representations is a brave (may be too brave) extrapolation of our meagre knowledge of strong interaction dynamics, almost all of which comes from *QCD*.
Well what about *QCD*? Does it fit in this scheme? Can we at least explain the known chiral symmetry breaking in *QCD*? The answer is yes. In *QCD* the quarks transform with the representation $3 \oplus \bar{3}$. (The reader should remember that the quarks and the charge conjugated anti-quarks are all put into left handed Weyl multiplets). In particular if we look at the di-fermion bound states we see that $(3 \oplus \bar{3}) \otimes (3 \oplus \bar{3})$ $(\bar{3} \oplus 6 \oplus 8 \oplus 1) \oplus (8 \oplus 1 \oplus 3 \oplus \bar{6})$, and it contains two singlet channels. These singlet channels must be the most strongly bound ones, since phenomenologically we know that the color group $SU(3)$ is not broken. We will return to the chiral breaking of *QCD* shortly.
Our first tumbling hypothesis, however, is too weak to make predictions with. All it does is to guarantee that a condensate will form in the most strongly bound scalar channel. But it doesn't tell us which is the most strongly bound channel. Since we don't know the answer to this question, we have to make a second assumption, or alternatively, solve the strong interaction dynamics for an arbitrary group G. Needless to say, facing the formidable (if not impossible) alternative, Raby, Susskind and Dimopoulos introduced the following educated (eminently so) guess as the second piece of the tumbling hypothesis [*31*]:

Hypothesis 2

Given a gauge group G and its fermion representation D, the most strongly bound scalar multiplet is the di-fermion channel for which the value of the function

$$C_2(D_{1,2}) - C_2(D_1) - C_2(D_2)$$

is a minimum, where C_2 is the quadratic Casimir operator and D_1 and D_2 are any of the irreducible components of D, and $D_{1,2}$ is any of the irreducible components of $D_1 \otimes D_2$.

Let us first see how Raby, Susskind and Dimopoulos arrived at this criterion, which they appropriately named maximally attractive (one gluon exchange) channel (*MAC*) principle.
Let us approximate the potential between two fermions by one gluon exchange. This potential will be in the form

$$V(r) \sim \frac{g^2}{r} \lambda_a(1) \cdot \lambda_a(2), \tag{4.1}$$

n analogy with the Coulomb potential. This can be written as

$$V(r) \sim \frac{g^2}{2r} \left((\lambda(1) + \lambda(2))^2 - (\lambda(1)^2 - \lambda(2)^2)\right) = \frac{g^2}{2r} \left(C_2(1, 2) - C_2(1) - C_2(2)\right) \tag{4.2}$$

thus proving the hypothesis to the first order. Of course we stated the *MAC* criterion as a hypothesis, since for strong interactions, one can not assume in general that a first order result will not be spoiled by higher order corrections. Indeed, the higher order corrections may change the result drastically enough that a different channel becomes the *MAC* when all such effects are considered.

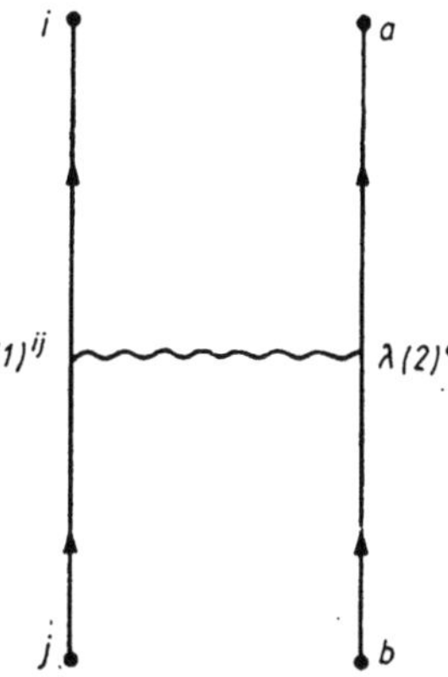

Fig. 4.1

This hypothesis (the *MAC* hypothesis) however, is probably on a stronger ground than it may appear at first sight, since some of the higher order corrections are taken into account when we use (as we have to) the running coupling constant $g(\mu^2)$. This single gluon exchange is a good approximation for small r (due to the asymptotic freedom); but with the running coupling constant replacing the constant g, this is probably a reasonable approximation to even much larger length scales. When it gives clear cut results we can perhaps rely on the *MAC* hypothesis with a reasonable assurance; however, when there are two or more different channels with comparable binding strength in the one gluon exchange approximation, the neglect of the higher order corrections starts manifesting itself as more and more important. Therefore the *MAC* hypothesis must be used with extreme caution, especially in those cases where this criterion favors one channel over the others by a small amount.
As a simple exercise, let's go back to the *QCD* example and study the *DSB* pattern under the light of the *MAC* hypothesis now:

$$qq \text{ condensate } 3 \otimes 3 = \bar{3} \oplus 6$$

$$\bar{q}\bar{q} \text{ condensate } \bar{3} \otimes \bar{3} = 3 \oplus \bar{6}$$

$$q\bar{q} \text{ condensate } 3 \otimes \bar{3} = 1 \oplus 8.$$

The second order Casimir invariants for $SU(3)$ are given as $C_2(3) = C_2(\bar{3}) = 4/3$, $C_2(6) = C_2(\bar{6}) = 10/3$, and $C_2(8) = 3$. Since we don't have to look at $\bar{q}q$ channel (it is just the conjugate of $q\bar{q}$) there are 4 possibilities:

$$qq\text{-antitriplet: } V \sim \frac{1}{2}\frac{g^2}{r}\left(C_2(\bar{3}) - C(3) - C(3)\right) = \frac{g^2}{2r}\left(-\frac{4}{3}\right)$$

$$qq\text{-sextet: } \quad V \sim \frac{1}{2}\frac{g^2}{r}\left(C_2(6) - C(3) - C(3)\right) = \frac{g^2}{2r}\left(\frac{2}{3}\right)$$

$$q\bar{q}\text{-octet:} \qquad V \sim \frac{1}{2}\frac{g^2}{r}\left(C_2(8) - C(3) - C(\bar{3})\right) = \frac{g^2}{2r}\left(\frac{1}{3}\right)$$

$$q\bar{q}\text{-singlet:} \qquad V \sim \frac{1}{2}\frac{g^2}{r}\left(C_2(1) - C(3) - C(\bar{3})\right) = \frac{g^2}{2r}\left(-\frac{8}{3}\right). \tag{4.3}$$

Of these, clearly the $q\bar{q}$ singlet is the MAC. Therefore the $SU(3)$ group itself cannot be broken, but any chiral flavor symmetry associated with it will be. We know all about this by now! Actually the MAC hypothesis answers another question we should have been asking long before, when we were building TC models, but tacitly avoided it: Why is it so in TC theories, that the TC group itself as always unbroken! The MAC principle answers this very easily, and we will state that result as a simple theorem:

Theorem: If the fermion representation D is real then the MAC is a singlet.

Proof: For any representation L of G, we have $C_2(L) > 0$ if L is not the singlet representation, and $C_2(1) = 0$. Let D_1 be an irreducible component of D. Since D is real, then either D_1 is real, or else $\bar{D}_1$ is also an irreducible component of D. In the first case C_2 (composite) $- C_2(D_1) - C_2(D_1)$ is minimum for C_2 (composite) $= 0$. In the second case C_2 (composite) $- C_2(D_1) - C_2(\bar{D}_1)$ is minimum again for C_2 (composite) $= 0$, implying a singlet.
Now it remains to show that a channel in $D_1 \otimes D_2$ cannot be the MAC, when $D_2 \neq D_1$ or $D_2 \neq \bar{D}_1$. To see this, assume the contrary, that the MAC is in $D_1 \otimes D_2$ where $D_1 \neq D_2$ and $D_2 \neq \bar{D}_1$. In this case $D_1 \otimes D_2$ cannot be a singlet, and $V \sim g^2/2r$ (C_2(comp.) $- C_2(D_1) - C_2(D_2)$). Without loss of generality assume $C_2(D_1) > C_2(D_2)$. In this case the singlet in the $D_1 \otimes D_1$ (or $D_1 \otimes \bar{D}_1$ if D_1 is not real) channel will have $V \sim g^2/2r \left(C_2(1) - C_2(D_1) - C_2(D_1)\right)$ which is more negative, and therefore it contradicts the fact that the MAC was in the $D_1 \otimes D_2$ channel, thus proving the theorem.

As an application for this theorem, in QCD, or in general in all vector-like theories the condensate is guaranteed to be a singlet. This is why in the TC theories the gauged group didn't break.

Let us emphasize two points. First of all, when the MAC hypothesis is used, it is assumed that the MAC will be a di-fermion bound state, not a multi-fermion bound state. Clearly QCD supports this hypothesis; the most strongly bound states are the lightest pseudo-scalar-octet members, which seem to be predominantly $q\bar{q}$ bound states, with very little (if any) $qq\bar{q}\bar{q}, \ldots$ mixtures. Nevertheless, one must not forget the special place QCD occupies amongst the gauge theories. It won't be surprising, if this property of QCD was shared by all vector-like theories; but *assuming that the MAC will be in a di-fermion channel in an arbitrary gauge theory with a non-real fermion representation content may be too strong an assumption.* The second point we want to emphasize is that the *MAC hypothesis determines the condensate, but it does not determine the direction of the symmetry breaking, except in some simple cases.* We shall return to this case later, and mention several attempts so solve this problem. We will also give an explicit example demonstrating that.
The reader might ask at this point why we restrict the condensates to scalars and/or pseudo scalars by the above two hypotheses. Why not a vector condensate for example? The obvious answer one comes up with is Lorentz invariance. At first sight it appears that such a condensate will violate the Lorentz invariance, therefore one can claim that it cannot be! But then, how does the condensate know it is not supposed to break the Lorentz invariance? The only rule is that the most strongly bound multiplet should condensate, and in principle, if the most strongly bound multiplet happens not to be a

singlet under the proper Lorentz group, it still should condensate, thus perhaps breaking the Lorentz group.

The original tumbling hypotheses by Raby, Susskind and Dimopoulos can not satisfactorily answer (and do not attempt either) this problem. *It is assumed that the condensate will be a scalar.* Fortunately enough, this assumption seems to work with the chiral breaking of *QCD*. In this case the most tightly bound (hence the lightest) bound states are the members of the pseudo-scalar octet, all of which we call "pions" in a generalize sense.

There have been recent attempts [34] based on an observation by Bjorken [35] long before the advent of quantized gauge theories. These attempts center around the possibility that the condensate could very well be a vector condensate, thus violating the Lorentz invariance of the vacuum, but according to Bjorken's explanation [35] such a violation may be impossible to observe. Since this review is intended to be at an introductory level, we will not attempts to elaborate this point any further in order not to leave the mainstream of the ideas of tumbling. We refer the interested reader to the original literature.

Before proceeding any further, let us give a simple example due to Georgi originally [36]. For $SU(n)$, we'll denote the completely antisymmetric product of m fundamental representations by $[m]_n$

$[m]_n$ =  m squares

$$= (0, 0, \ldots, 0, m, 0, \ldots, 0) = (a_1, \ldots, a_{n-1}),$$
$$= (1, 1, \ldots, 1) = (\mu_1, \mu_2, \ldots, \mu_m),$$

where the notation in terms a_i or μ_j is defined in the Appendix C. The ABJ triangle anomaly associated with these anti-symmetric representations is easily computed [36] to be

$$A([m]_n) = \frac{(n-2m)\,(n-3)!}{(n-m-1)!\,(m-1)!}. \tag{4.4}$$

We see then immediately that the following reducible representation of $SU(2n+1)$

$$[2]_{2n+1} \oplus [4]_{2n+1} \oplus \cdots \oplus [2n]_{2n+1}$$

is anomaly-free. Equipped with this knowledge, we introduce an $SU(5)$ gauge theory with the following anomaly-free fermion content:

$$[2]_5 \oplus [4]_5 = 10 \oplus \bar{5}.$$

ψ_{ab} and χ_{abcd} are the left handed Weyl multiplets of 10 and $\bar{5}$ respectively, and the indices $a, b, c, d, e, \ldots$ run over 1 to 5 and they are completely anti-symmetrized on ψ and χ. Let's find the one gluon exchange potential for the di-fermion condensates:

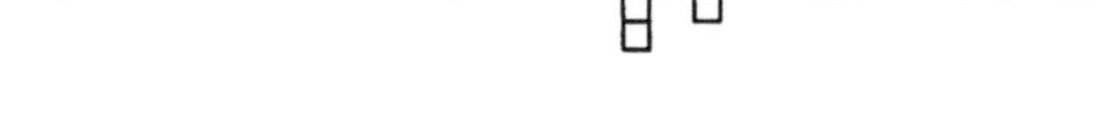

Using (C.1) given in the Appendix C for $SU(n)$, we write out the second order Casimir invariants for the representations involved as follows:

$$
\begin{aligned}
C_2(5) &= C_2(\bar{5}) = 12/5 \\
C_2(10) &= C_2(\overline{10}) = 18/5 \\
C_2(15) &= C_2(\overline{15}) = 28/5 \\
C_2(45) &= C_2(\overline{45}) + 32/5 \\
C_2(50) &= C_2(\overline{50}) = 42/5 .
\end{aligned}
\tag{4.5}
$$

Then we can tabulate the results:

Bound state	Cross product	$2rV/g^2$
50		$6/5$
45	$10 \otimes 10$	$-4/5$
$\bar{5}$		$-24/5 \leftarrow MAC$
$\overline{45}$	$10 \otimes \bar{5}$	$2/5$
5		$-18/5$
$\overline{15}$	$\bar{5} \otimes \bar{5}$	$4/5$
$\overline{10}$		$-6/5.$

Therefore we see that the $\bar{5}$ multiplet coming from $10 \otimes 10$ will condensate. The condensate can be written as

$$\varphi_e = \psi_{ab}\psi_{cd}\varepsilon_{abcde} \, .$$

By an $SU(5)$ rotation we can choose $\langle \varphi_e \rangle = 0$ for $e = 1, 2, 3, 4$, and $\langle \varphi_5 \rangle \neq 0$. Therefore the symmetry will be broken to $SU(4)$, and the 9 generators in the coset space $SU(5)/SU(4)$ will gain masses of order μ, where μ is computed from

$$\frac{g^2(\mu)}{4\pi} \frac{24}{5} \sim 1, \tag{4.6}$$

as an order of magnitude estimate if $g^2(\mu)$ is known. The fermion representations of $SU(5)$ decompose under the unbroken $SU(4)$ group as follows:

$$
\begin{aligned}
[2]_5 &\to [2]_4 \oplus [1]_4 = 6 \oplus 4 \\
[4]_5 &\to [4]_4 \oplus [3]_4 = 1 \oplus \bar{4} .
\end{aligned}
$$

The condensate $\bar{5}$ of $SU(5)$ is formed in the $10 \otimes 10$ product must be an $SU(4)$ singlet, therefore it is necessarily formed in the singlet channel coming from $6 \otimes 6$ of $SU(4)$. (We remind the reader that $[2]_4 = 6$ is a self adjoint representation of $SU(4)$, therefore $6 \otimes 6$ does contain a singlet!) Therefore we conclude that the fermions in $[2]_4$ gain masses of order μ, and are eliminated from the effective Lagrangian describing physics at $\mathcal{E} \ll \mu$. At such a scale, we have an effective $SU(4)$ gauge theory with the following fermion content:

$$[0]_4 \oplus [1]_4 \oplus [3]_4 = 1 \oplus 4 \oplus \bar{4} .$$

It can easily be checked that this representation is anomaly-free. (As a matter of fact it is real). Again we can work out the relevant direct products

$$4 \otimes 4 = 6 \oplus 10$$
$$4 \otimes \bar{4} = 1 \oplus 15$$
$$\bar{4} \otimes \bar{4} = \bar{6} \oplus \overline{10}$$

and calculate the Casimir invariants from (C.1) in the appendix

$$\begin{aligned} C_2(1) &= 0 \\ C_2(4) &= C_2(4) = 15/8 \\ C_2(6) &= 5/2 \\ C_2(10) &= C_2(10) = 9/2. \end{aligned} \tag{4.7}$$

Then the MAC is found to be in the singlet channel coming from $4 \otimes \bar{4}$. Thus the group $SU(4)$ remains unbroken. The fermion multiplets $[1]_4$ and $[3]_4$ gain dynamical masses of order μ' (and are expected to be confined). In principle if μ is known, μ' can be computed from the β function.

After this stage the only uncondensed fermion left is $[0]_4$, the singlet. But the singlet cannot be effected by any further increase of the $SU(4)$ coupling constant, and the theory does not tumble any more.

So this is a simple example where the theory tumbles once (from $SU(5)$ to $SU(4)$) but two distinct mass scales appear in it (μ, μ').

Next, we want to give another example and demonstrate with it one of the shortcomings of the two tumbling hypotheses:

Let us take an $SU(7)$ gauge theory with the following fermion content

$$[6]_7 \oplus [6]_7 \oplus [3]_7 = \bar{7} \oplus \bar{7} + 35,$$

which can be checked immediately to be anomaly-free. In this case the relevant cross products are as follows: (we strongly urge the readers with little experience in this sort of thing to work these out explicitly using Young tableaus.)

$$\bar{7} \otimes \bar{7} = \overline{21} \oplus 28$$
$$\bar{7} \otimes 35 = 21 \oplus \overline{224},$$
$$35 \oplus 35 = 588 \oplus 490 \oplus 140 \oplus \bar{7}.$$

The relevant Casimir invariants are:

$$\begin{aligned} C_2(\bar{7}) &= 24/7 \\ C_2(21) &= C_2(\overline{21}) = 40/7 \\ C_2(35) &= 48/7 \\ C_2(140) &= 66/7 \\ C_2(\overline{224}) &= 75/7 \\ C_2(490) &= 108/7 \\ C_2(588) &= 94/7. \end{aligned} \tag{4.8}$$

Then we conclude that the MAC has to be the 7 channel in the $35 \otimes 35$ product. The observant readers have probably noticed that until now the MAC has always been the smallest irreducible component in the largest (in dimension) direct product. Such need not be the case in all applications; ambiguities occur often, as we shall see. However it is usually safe to check several small components of larger direct products only. For example without computing any Casimir operators, we could have safely eliminated the 490 and 588 of $35 \otimes 35$, and $\overline{224}$ of $\bar{7} \otimes 35$, and 28 of $\bar{7} \otimes \bar{7}$ from consideration.
Something new happens here, however. Let's write the condensate as follows

$$\varphi_g = \varepsilon_{abcdefg} \chi_{abc} \chi_{def} = 0, \text{ because} \qquad (4.9)$$
$$[\chi_{abc}, \chi_{def}]_+ = 0^{20)}.$$

Now we have to check the other channels for the MAC. Using the C_2 values given above, we conclude that 21 piece of the product $35 \otimes \bar{7}$ is the MAC in the absence of $\bar{7}$ of $35 \otimes 35$. But now, we are stuck! How does the 21 break the symmetry? So far the two hypotheses of tumbling tell us what the condensate is, but they give no information about the direction of symmetry breaking. Without any further input at this moment, we are unable to decide how this $SU(7)$ gauge theory tumbles. When the condensate transforms with one of the fundamental representations (in the present case this will be a 7 or a $\bar{7}$) such a problem does not arise; but with 21 or $\overline{21}$, we reed the extra input. Of course, the problem of finding the pattern of symmetry breaking was studied long ago in relation to Higgs scalars, and some simpler representations of the classical Lie groups were worked out in detail by Ling-Fong Li [*37*], and V. Elias, S. Eliezer, A. R. Swift [*38*]. At this point Raby, Susskind and Dimopoulos suggest the use of Li's analysis to decide about the pattern of DSB. If we follow this line of thought we have to introduce the following third hypothesis:

Hypothesis 3

To decide about the direction of the symmetry breaking, treat the condensate as if it is an elementary scalar field with an effective Lagrangian containing no higher than fourth order polynomial terms. Then minimize the scalar potential, just as it is done in the Higgs mechanism.

This hypothesis too, is in general not always sufficient to solve all the ambiguities in the determination of the direction of DSB. Some residual ambiguity may remain due to the fact that the effective (fourth order) potential has some number (two or more) of parameters which are completely arbitrary in a truly elementary scalar field theory, and the interplay of the relative signs and sizes of these parameters changes the direction of spontaneous symmetry breaking. Of course in a dynamically broken theory, one hopes to be able to calculate these self-coupling parameters from the knowledge of the strong interaction dynamics. Even if this true in principle, clearly we are not equipped with such tools to compute these parameters yet.
Let us go back to our example of $SU(7)$ and study it further now. 21 is the twice anti-symmetric product of two 7's. The symmetry breaking pattern of $SU(n)$ for the second order anti-symmetric tensor scalar multiplet was computed by Li [*37*] and the result was found to be $SU(n-2) \otimes SU(2)$, or $Sp(2[n/2])$, where the notation $[x]$ is used for the

[20]) It's obvious that in general the Fermi statistics prohibit any condensate which appears in an anti-symmetric product of a representation by itself, if the representation in question is $(2m+1)$-fold antisymmetric product of the fundamental representation of $SU(n)$. $(2m+1 \leqq n/2)$. For example the 11 of $SU(11)$ appearing in the product $462 \otimes 462$ can not condensate for the same reason.

largest integer smaller or equal to x. In general for the second order anti-symmetric tensor there are two quartic terms in the potential and which path of symmetry breaking is followed is determined by them.
So in our example of $SU(7)$, the unbroken group will either be $SU(5) \otimes SU(2)$, or $Sp(6)$ and we have no way of deciding on that yet.
Recently there have been attempts modifying (or supplementing) this third hypothesis to resolve the ambiguities when they appear [*39*].
The observant reader has by now noticed that the above three assumptions of tumbling are given according to the order of their likely reliability. In our opinion, it is very difficult to find any strong argument against the first hypothesis. The second hypothesis, we believe might give a fair enough of an approximation for the majority of the cases; however we won't be surprised if it fails in some cases, even as an approximate answer. The third hypothesis is the shakiest of them all, and it is conceivable (even likely) that it usually gives a wrong answer. This weakness of the third hypothesis stems from the fact that the scalar potential in question which solely determines the direction of symmetry breaking is restricted to be a fourth order polynomial. For the elementary scalar fields this restriction is the direct result of the renormalizability. However for a set of composite scalars the situation is different, the renormalizability puts no restrictions on the order of the scalar potential, which in fact could be (and should be) of infinite order. Other alternatives to the third hypothesis using this observation as a starting point habe been sought recently. An interesting one of these [*40*] allows a scalar potential of infinite order, and assumes that it can be approximated by one-fermion-loop effective coupling. In the future; as we understand more about the *DSB*, no doubt *better constructed rules* (*or hypotheses*) *will replace the third hypothesis* of Raby, Susskind and Dimopoulos. We see it as very likely that *even the second hypothesis will be modified, if not replaced.*
Before we go on to the next part where we will work out a particular tumbling pattern in detail, let us ask one other question not addressed yet: When does the tumbling stop? One of the cases where it has to stop was encountered in the $SU(5)$ example we presented. $SU(5)$ tumbles down to $SU(4)$ once, generating two mass scales and it stops because the only fermion left after this stage, which has not yet participated in any condensate (therefore has not yet acquired a dynamical mass) is the $SU(4)$ singlet, which is blind to the development of the $SU(4)$ coupling constant any further. This situation is generally true, of course; whenever nothing but a set of singlets of fermions are left under the last unbroken group, the tumbling has no choice but to stop there. A second possibility exists, even though for obvious reasons, any theory which hopes to be even semi-realistic cannot take advantage of it. It arises when the last unbroken subgroup reduces to a direct product of $U(1)$ groups, thus being abelian. Its coupling constants can no longer increase with the decreasing mass scale and can not generate another round of *DSB*. Since we want at least the color $SU(3)$ to be a subgroup of the unbroken local symmetries of nature, we don't expect to encounter this possibility in any phenomenologically interesting model. A third possibility, also unphysical, is that all fermions may participate in condensates and at a certain stage no massless fermions may be left, which again forces the tumbling mechanism to stop. Clearly there is no room for leptons in such a theory, since all the participating fermions are confined by the strong binding forces.

c) A Detailed Example

We would like to present an example now, which tumbles several times, generates different mass scales and different families of fermions. This is an unrealistic toy model of tumbling, originally due to Raby, Susskind and Dimopoulos [*31*], but it exhibits all the novel features of the theory. We start with an $SU(8)$ theory[21]) with the following fermion content:

$$[7]_8 \oplus [6]_8 \oplus [3]_8 = \bar{8} \oplus \overline{28} \oplus 56.$$

Using the anomaly formula (4.4) the reader can verify that this is an anomaly-free representation. The various cross products in this model are:

$$\begin{aligned} \bar{8} \otimes \bar{8} &= \overline{28} \oplus \overline{36} \\ \bar{8} \otimes \overline{28} &= \overline{56} \oplus \overline{168} \\ \bar{8} \otimes 56 &= 28 \oplus \overline{420} \\ \overline{28} \otimes \overline{28} &= \overline{70} \oplus \overline{336} \oplus \overline{378} \\ \overline{28} \otimes 56 &= 8 \oplus \overline{216} \oplus \overline{1344} \\ 56 \otimes 56 &= \overline{28} \oplus 420 \oplus 1176 \oplus 1512. \end{aligned} \tag{4.10}$$

If the reader remembers our comment immediately following eq. (4.8) about finding the MAC without computing all C_2 values and $V(g)$ potentials, he can convince himself at once, that the only channels which are candidates for the MAC are $\overline{28}$ of $56 \otimes 56$, 8 of $\overline{28} \otimes 56$, 70 of $\overline{28} \otimes \overline{28}$, and 28 of $\bar{8} \otimes 56$. The other channels clearly won't make it. The C_2 values for the representations involved are

$$\begin{aligned} C_2(8) &= C_2(\bar{8}) = 63/16 \\ C_2(28) &= C_2(\overline{28}) = 108/16 \\ C_2(56) &= C_2(\overline{56}) = 135/16 \\ &\quad\; C_2(\overline{70}) = 144/16 \end{aligned} \tag{4.11}$$

which immediately give 8 of $\overline{28} \otimes 56$ as the MAC when substituted into (4.2). Since 8 is one of the fundamental representations of $SU(8)$, we know immediately that it breaks the symmetry to $SU(7)$ at a mass scale which we call $m_{(1)}$. 15 gauge bosons of $SU(8)/SU(7)$ gain masses of order $m_{(1)}$. To decide what happens to the fermions we decompose the fermion representation of $SU(8)$ under $SU(7)$:

$$\begin{aligned} [7]_8 &\to [6]_7 \oplus [7]_7 = \bar{7} \oplus 1 \\ [6]_8 &\to [5]_7 \oplus [6]_7 = \overline{21} \oplus \bar{7} \\ [3]_8 &\to [2]_7 \oplus [3]_7 = 21 \oplus 35. \end{aligned}$$

Since the MAC is in the $[6]_8 \otimes [3]_8$ product and since it is an $SU(7)$ singlet, necessarily it must come from $[2]_7 \otimes [5]_7$ piece under $SU(7)$ ($21 \otimes \overline{21}$ is the only possible $SU(7)$ product which contains an $SU(7)$ singlet). Therefore we conclude that $[2]_7$, and $[5]_7$ multiplets gain masses of order $m_{(1)}$.
At mass scales $\mu \ll m_{(1)}$ the effective theory is an $SU(7)$ theory with the following fermion content:

$$[7]_7 \oplus [6]_7 \oplus [6]_7 \oplus [3]_7 = 1 \oplus \bar{7} \oplus \bar{7} \oplus 35.$$

Other than the singlet $[7]_7$, this effective $SU(7)$ theory is identical to the one we considered in section (b) of this part. The reader will remember that since the expected MAC $\bar{7}$ of

[21]) The model we present here is actually simpler than that of reference [*31*]. This model can be embedded in the $SU(9)$ model of reference [*31*].

$35 \otimes 35$ vanished (see equation (4.9)) due to the Fermi statistics, the MAC was the 21 of $\bar{7} \otimes 35$. How does 21 of $\bar{7} \otimes 35$ break the $SU(7)$ symmetry? It was exactly this question that led us to the introduction of the third hypothesis of tumbling, so let's make use of it. Using LI's analysis [37] we conclude that the breaking can procede in two different directions: $SU(7) \sim Sp(6)$, or $SU(7) \to SU(5) \otimes SU(2)$ depending upon the relative sizes and signs of the effective quartic coupling of the bound state scalars. We cannot decide yet which one it is! So let's take a look at them both.
But first let's point out that the symmetry of this model is not quite $SU(7)$. There is also an accidental global flavor symmetry due to the presence of two $[6]_7$ multiplets of fermions. This global $SU(2)$ symmetry is broken by the condensate (21 breaks this global $SU(2)$ completely), this we will have 3 NGB's corresponding to the broken generators of the global $SU(2)$ group. We don't have to worry about the masslessness of these however. Remember that one of the $[6]_7$ multiplets came from $[7]_8$ of the broken $SU(8)$, the other from $[6]_8$. Therefore under the action of the (now very massive) $SU(8)/SU(7)$ gauge bosons the two $[6]_7$ multiplets are quite different, and their degeneracy is lifted; in other words this global $SU(2)$ flavor symmetry is not only spontaneously broken by the condensate 21, but also explicitly broken by the massive $SU(8)/SU(7)$ gauge bosons. Here $SU(8)$ acts like an ETC group, and the generators of $SU(8)/SU(7)$ correspond to the massive ETC gauge bosons. Therefore, the three NGB's we find are actually PNGB's gaining one loop (or higher) masses from the $SU(8)$ interactions. These interactions in principal determine two linear orthogonal combinations of the original two $[6]_7$ multiplets, say $[6']_7$ and $[6'']_7$, and the condensate $[2]_7$ occurs in the cross product $[6']_7 \otimes [3]_7$, and not in $[6'']_7 \otimes [3]_7$. Without this explicit breaking, both $[6]_7$ multiplets would have participated in the condensate[22]).
Now back to the tumbling patterns. Let us look at the $Sp(6)$ channel first. Since all representations of $Sp(6)$ are real, each cross product with itself will contain a singlet and one of these singlets will be the MAC. The fermions which participate in the singlet will gain masses of order $m_{(3)}$, but the group $Sp(6)$ will not be broken. At a later stage (but still not far from $m_{(3)}$) the next singlet will condensate, still preserving the $Sp(6)$ symmetry. Since each representation can give a singlet when we take the cross product with itself, by repeating the above process several times all the fermions will gain masses of order $m_{(3)}$ (not quite degenerate though, some of this degeneracy is lifted by the several stages of condensation happening at lower mass scales, but not an order of magnitude lower). The only fermions which can escape receiving dynamical masses are the $Sp(6)$ singlets, and there are three of those, since $[6]_7$ of $SU(7)$ decomposes under $Sp(6)$ as $6 \oplus 1$. So this branch of tumbling stops at $Sp(6)$ having consumed all of its non-singlet fermions. The Table 4.1 summarizes this branch of tumbling, where $1', 1'', 6'$, and $6''$ label the $Sp(6)$ contents of the $SU(7)$ representations $[6']_7$ and $[6'']_7$.
Next let's study the tumbling pattern for the $SU(5) \otimes SU(2)$ channel which will prove to be much richer in structure. Again we start by writing the fermion content of the effective theory:

$$[6']_7 \to [4', 0]_{5,2} \oplus [5, 1']_{5,2} = (\bar{5}', 1) \oplus (1, 2')$$

$$[6'']_7 \to [4'', 0]_{5,2} \oplus [5, 1'']_{5,2} = (\bar{5}'', 1) \oplus (1, 2'')$$

$$[3]_7 \to [3, 0]_{5,2} \oplus [2, 1]_{5,2} \oplus [1, 0]_{5,2} = (\overline{10}, 1) \oplus (10, 2) \oplus (5, 1).$$

---

[22]) There is of course the unknown, whether or not the breaking by the heavy $SU(8)/SU(7)$ loop interactions is strong enough to make an important difference in the binding strengths of $[5']_7$ and $[5'']_7$ coming from $[1']_7 \otimes [4]_7$ and $[1'']_7 \otimes [4]_7$ respectively. If these are close enough, both may condensate thus creating problems for the next stage of tumbling for the $SU(5) \otimes SU(2)$ channel, since the effective fermion content of $SU(5) \otimes SU(2)$ will no longer be anomaly free, without the inclusion of heavier fermions.

Table 4.1

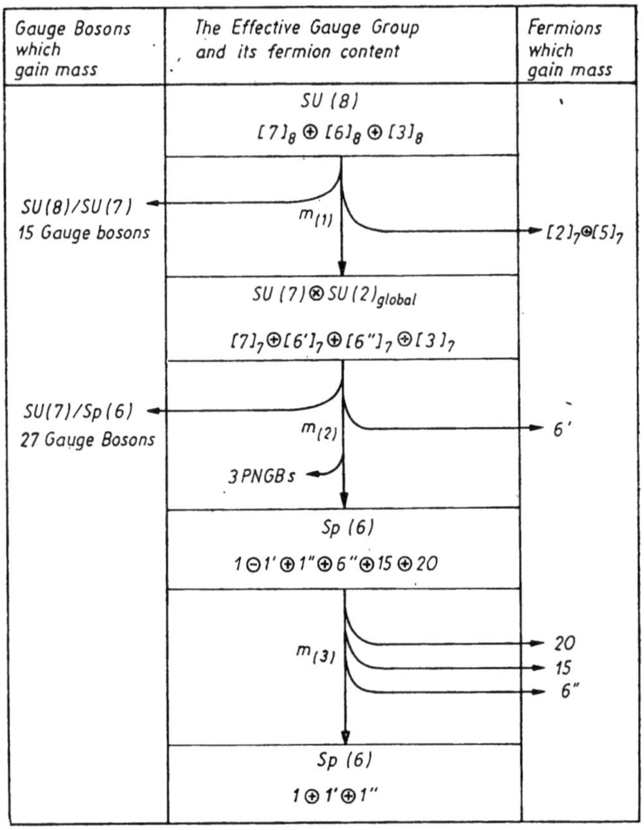

(read ⊕ instead of ⊖)

Since the condensate is the $SU(5) \otimes SU(2)$ singlet in the product $[6']_7 \otimes [3]_7$, the only candidate for it is the singlet in the $[4', 0]_{5,2} \otimes [1, 0]_{5,2}$ direct product. Thus we see that at a mass scale of $m_{(2)}$ the fermion multiplets $(\bar{5}', 1)$ and $(5, 1)$ and the 21 gauge bosons of $SU(7)/(SU(5) \otimes SU(2))$ gain masses roughly the same order as this mass scale.
At energies much below $m_{(2)}$ the effective theory is an $SU(5) \otimes SU(2)$ theory with the following fermion content:

$$[0, 1']_{5,2} \oplus [4'', 0]_{5,2} \oplus [0, 1'']_{5,2} \oplus [2, 1]_{5,2} \oplus [3, 0]_{5,2}.$$

The β function of the $SU(5)$ group, however, grows much faster than that of the $SU(2)$ group, and for the purpose of the next stage of *DSB*, the $SU(2)$ group acts as a bystander. In that case, without even doing any computation of the effective potential, we can say at once, that the *MAC* will be the $SU(5)$ singlet in the product $(10, 2) \otimes (\overline{10}, 1) = (1, 2) \oplus (99, 2)$. Clearly this condensate does not break $SU(5)$ but it breaks $SU(2)$ being a doublet under it. The doublet breaks this $SU(2)$ completely (no $U(1)$ subgroup can survive this symmetry breaking), hence the unbroken group to set the next stage will be $SU(5)$

only. The fermion representation decomposes with respect to $SU(5)$ as follows:

$$(1, 2') \to 1 \oplus 1$$
$$(5'', 1) \to 5$$
$$(1, 2'') \to 1 \oplus 1$$
$$(10, 2) \to 10 \oplus 10$$
$$(\overline{10}, 1) \to \overline{10}.$$

The $SU(5)$ singlet then, will be formed in the product $10' \otimes \overline{10}$. Note that $10'$ and $10''$ again are going to be orthogonal linear combinations of the two 10. The linear combination which participates in the condensate is decided by the explicit lifting of the degeneracy by the $SU(2)$ group[23]). We conclude then, the fermions $10'$, $\overline{10}$ of $SU(5)$ and the 3 gauge bosons of $(SU(5) \otimes SU(2))/SU(5)$ gain masses of order $m_{(3)}$.
At energy scales much below $m_{(3)}$ the effective theory is an $SU(5)$ theory with the following fermion content

$$1 \oplus 1 \oplus 1 \oplus 1 \oplus \bar{5} \oplus 10''.$$

Other than the four extra singlets (which don't effect anything anyway) this theory is identical to the one we studied in the previous section. It tumbles down to $SU(4)$ once more, and generates two more mass scales, say $m_{(4)}$ and $m_{(5)}$. So the whole tumbling pattern of the $SU(5) \otimes SU(2)$ channel is as mapped in Table 4.2.
Well, what is the moral lesson that we are supposed to learn from this toy model? First of all we see that we can generate many different fermion mass scales; we can also generate many sequential, strong, asymptotically free groups that play the roles of TC, TC', ... Furthermore, now we have a natural hierarchy of effective ETC groups and effective $TC, TC', \ldots$ groups, and in principle we can even relate the coupling constants of two strong groups to each other. In the above model, there are two arbitrary parameters: g of $SU(8)$ group and one mass scale. All other mass scales, $m_{(1)}, \ldots, m_{(5)}$, as well as coupling constants can in principle (in reality an extremely difficult task) be calculated.

d) Concluding Remarks and Future Prospects

Let us go back to the GUT's with elementary scalars again, and see to what problems of these theories we can offer a "cure" now, through the combination of our experience with *DSB*, technicolor and tumbling. We would like to emphasize that this "cure", is as of yet, "a cure in principle". We have to tell the reader immediately, that to this date, no dynamically broken quasi-realistic model has been known to us. There were a number of attempts by several authors [*26, 41, 43*] in that direction, but these models have not penetrated the barrier of being toy models. With this warning in mind, we can start listing our "cures" now:

1. Since we no longer have any elementary scalars at all, *the dynamically broken GUT's will not suffer from the loss of asymptotic freedom,* as long as the fermion content is not unreasonably large.
2. *No parameters will have to be adjusted to absurd accuracies to make the theory work!*

---

[23]) Since the $SU(2)$ coupling is weak compared to $SU(5)$ coupling, we expect the singlets in the products $10' \otimes \overline{10}$ and $10'' \otimes \overline{10}$ to have fairly close binding energies. There is the possibility, then, that they may both condensate, creating the same problem mentioned in footnote 22. In these situations it is probably not safe to rely on the second and the third hypotheses.

Table 4.2

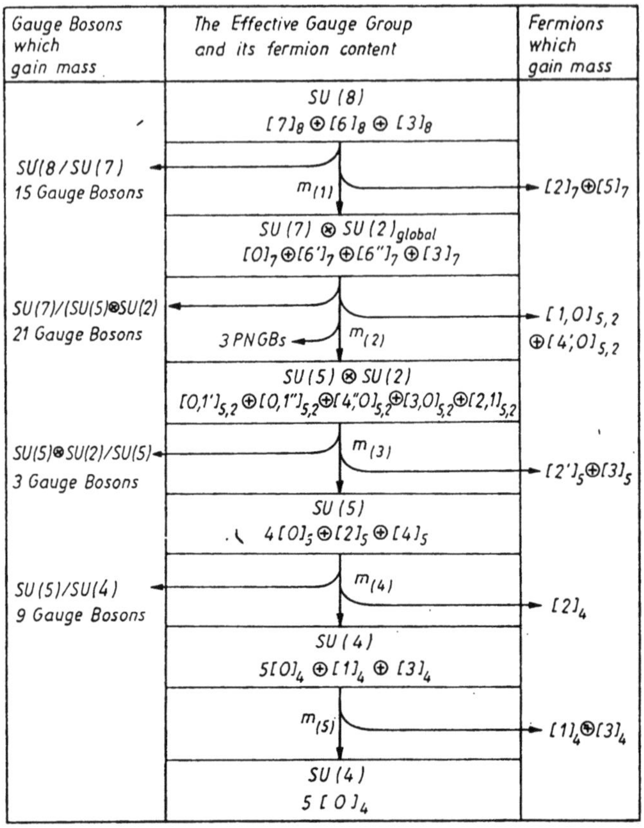

3. The two mass scales of GUT's, one at 300 GeV, the other at 10^{15} GEV need no longer be put there by hand. These scales will be generated spontaneously, along with several more intermediate ones. *The DSB will fill the gap between* 10^2 GeV *and* 10^{15} GeV *with very interesting physics, rich in structure and implications; the so called "desert" will bloom.*

4. *In principle every coupling constant, every mass, and every other physical quantity will be calculable in terms of only two arbitrary parameters.*

To any reader who followed these lectures, the first, second and the fourth comments must be obvious. The third one, perhaps needs a little further commenting: If it so happens that the prediction of the GUT's with elementary scalars is correct, and there are no intermediate mass scales in between 10^{15} and 10^2 GeV, the implications on the experimental high energy physics for the next century could be devastatingly bleak (The theorists don't fare much better either). Even if the money and the size of the job was no problem at all, even if we are given trillions of Dollars to build an accelerator as large as we can ever hope to build on Earth, namely one around the Equator, this still will not be enough for us to cross the "desert" and reach the next energy scale!
Of course, just because something looks unpromising and empty for future generations does not make it wrong. Despite the recent enthusiasm and the accelerating "coversion" rate to the belief that the symmetries must be broken dynamically, *there are still a lot of die-hard believers of the GUT's with elementary scalars.* But they too are working hard on those points we mentioned above; they too are unhappy about too much arbitrariness,

asymptotic freedom, the accuracy by which some parameters have to be adjusted, even about the great empty desert! They are trying to devise schemes to cure these points, at least some of them. It is just that they are not ready to give up their beloved elementary scalar fields yet. *Our (naturally strongly biased) view is that the theory of elementary scalars has consumed too much of our time and attention which might have been better spent on understanding more interesting alternative possibilities.*
Well, will we ever know the answer? The answer is not only yes, but possibly very soon. At the presently available energies in the accelerators we could see some of the lightest PNGB's, which are sometimes called technions. Most of these particles will be in the 100 GeV range and will not be observed until the next generating, but a few of them could be lighter than 20 GeV, (possible as light as 5 GeV), and a systematic search for these must soon be carried out. In these lectures we have not had any time to study the phenomenology of these technions, and we refer the reader to the literature [*12, 43—50*]. But the fact remains that the dynamically broken theories predict a whole host of scalars (perhaps more than a hundred) unter 500 GeV, while the gauge theories with elementary scalars do not predict many scalars in that energy range. (One, two, perhaps three). Again the reader is referred to the literature about the experimentally distinguishable characteristics of elementary and composite scalars. *Give credit to the dynamical symmetry breaking: at least it gives the experimentalists a shot at disproving it, unlike the theories with elementary scalars which can use the infinite freedom of endless adjustable parameters to run away when dangereously cornered!*

Appendix A

Maximal Flavor Preservation

Here we will reproduce an argument by Coleman and Witten [*51*], which shows that the maximal flavor symmetry is preserved in the *DSB* for large *N QCD*. This argument for Technicolor is less valnurable than *QDC*, because in the latter case on could always argue that in the actual world $N_c = 3$ (too far from the value where the argument holds); here we are less constained by physics. We will present the proof for the *SU*-case only. We first list the assumptions.

a) $G_{TC} = SU(N), \qquad N = N_{TC}$.
b) Techniquarks belong to the fundamental representation of G_{TC}.
c) Large-N limit exists.
d) G_{TC} yields confinement for the techni-quarks for arbitrarily large N.
e) If there exist n flavors, the *DSB* of $G_f = U(n)_L \otimes U(n)_R$ is characterized by a G_{TC}-invariant condensate of the form

$$\Sigma = \langle \varepsilon_{\alpha\beta} \bar{Q}^{ai}_{\alpha L} I^{ab}_{ij} Q^{bj}_{\beta R} \rangle \tag{A.1}$$

in the notation of Section 2.b. The chiral transformation properties of I in the 4-component language are

$$(U_L, U_R): I \to U_L{}^+ I U_R . \tag{A.2}$$

By a transformation of this form one can always make I real, diagonal and non-negative. The squares of the diagonal entries are the eigenvalues of the matrix I^+I. Let us recall that we have defined th evacuum alignment problem as the problem of finding the true vacuum which corresponds to the minimum of an effective potential V. For fixed embedding $G_w \subset G_f$, this V will be a function of these eigenvalues. Thus the pattern of *DSB* of $G_f \to H_f$ is determined by the pattern of the eigenvalues at the minimum of V. For example

a) If all the eigenvalues vanish at the minimum, there is no *DSB*.
b) If they are all equal, but non-zero, the symmetry breaks down to $U(n)$.
c) If they are all unequal and non-zero the symmetry breaks down to $U(1)^n$.

Under the assumptions listed above, breakdown beyond $U(1)^n$ is not possible. Now we state the claim we would like to prove as a theorem, for the SU-case.

Theorem: in the limit $N \to \infty$, chiral techniflavor group $G_f = U(n)_L \times U(n)_R$ necessarily breaks down to $H_f = U(n)_v$, in other words *DSB* pattern is such that maximal flavor symmetry is preserved.

Proof: If we expand $V[\Sigma]$ in powers of Σ we uncounter terms like $\mathrm{Tr}\,(II^+)^p$, $\mathrm{Tr}\,(II^+)^p \times \mathrm{Tr}\,(II^+)^q$, etc. Because traces of quark operators arise in Feynman diagrams from sums over fermion loops, the terms of the first sort come from diagram with 1-fermion loop, those of the second sort from diagrams with 2-fermion loops. etc. In the large-N limit the dominant diagrams are those with only 1-fermion loop [*52*], thus, to leading order in $1/N$, we have

$$V[\Sigma] \simeq N\,\mathrm{Tr}\,F[II^+] \tag{A.3}$$

where F is some N-independent function. Let us now denote the eigenvalues of II^+ by σ_i, $i = 1, \ldots, n$ then

$$V[\Sigma] \simeq \sum_i NF[\sigma_i]. \tag{A.4}$$

Since eigenvalues are all independent, to minimize this sum is to minimize each term separately. Each eigenvalue must be at the minimum of F:

$$\left.\frac{\partial F(\sigma_i)}{\partial \sigma_i}\right|_{\sigma_i=\bar{\sigma}_i} = 0. \tag{A.5}$$

Since all the $\bar{\sigma}_i$ correspond to the minimum of the same function, they are all equal:

$$\bar{\sigma}_1 = \bar{\sigma}_2 = \cdots = \bar{\sigma}_n. \tag{A.6}$$

This corresponds to $H_f = U(n)_v$ if $\bar{\sigma} \neq 0$. We shall next show that $\bar{\sigma} = 0$, that is no-symmetry-breaking case, is ruled out. Thus the symmetry necessarily breaks down, but in such way that the maximum flavor symmetry is preserved.
For this we shall consider a chiral current $j_\mu = \bar{\psi} A (1 + \gamma_5)\, \gamma_\mu \psi$ where A is an $n \times n$ Hermitian matrix. Now construct a three current correlation function:

$$\Gamma_{\mu\nu\lambda}(p, q, r) = \int\!\!\int d^4x\, d^4y e^{ip\cdot x} e^{iq\cdot y} T\langle j_\mu(x)\, j_\nu(y)\, j_\lambda(0)\rangle \tag{A.7}$$

where $r = -(p + q)$. It was Dolgov and Zakharov [*53*] who first noticed that the triangle graph associated with a current with an anomalous divergence cannot be analytic, say at $q^2 = 0$, using dispersion theoretics arguments. Here we shall briefly describe the simple permutation symmetry argument of Coleman and Witten. It is clear that Γ is symmetric under simultaneous permutations of (p, q, r) and (μ, ν, λ). The chromodynamic anomaly in the $U_A(1)$ current is proportional to $1/N$ and thus is irrelevant in the large-N limit [*54*]. Thus the Adler-Bell-Jackiw anomaly equation [*24*] states that

$$r^\lambda \Gamma_{\mu\nu\lambda} = \frac{N}{\pi^2}\,(\mathrm{Tr}\,A^3)\, \varepsilon_{\mu\nu\lambda\sigma} p^\lambda q^\sigma. \tag{A.8}$$

This equation implies that Γ cannot be analytic. To prove this suppose that Γ is analytic, that is it has a Taylor expansion. Thus the right hand-side of (A.8) must come from the first order term in this expansion. If it was not for the permutation symmetry of Γ, there would be two independent first order pseudotensors, $\varepsilon_{\mu\nu\lambda\sigma}p^\sigma$, and $\varepsilon_{\mu\nu\lambda\sigma}q^\sigma$. Permutation symmetry require that the first order term in the Taylor expansion be of the form

$$\Gamma_{\mu\nu\lambda}(p, q, r) \propto \varepsilon_{\mu\nu\lambda\sigma}(p + q + r)^\sigma = 0. \tag{A.9}$$

This completes the proof.

It was shown that in leading order in $1/N$ the only singularities in Green's functions made of strigs of quark bilinears are simple poles [52]. For Γ these poles occur at the values of p^2, q^2 and (or) r^2 equal to the masses of the particles created from the vacuum by the corresponding chiral currents. Because Γ is not analytic at $p = q = r = 0$, j_μ must create at least one massless particle from the vacuum. Since we are dealing with a massless field theory, Lorentz invariance tells us that only scalar particles can be created from the vacuum. This, then means that the symmetry corresponding to the conserved current j_μ is spontaneously broken. This completes the argument.

It is worth repeating that the most crucial assumption which went into the above proof is the strong group representation content of the techniquarks. For instance, if the techniquarks didnot belong to the fundamental representation, but instead transformed like a rank-2 tensor under G_{TC}, then the diagram with a 1-fermion loop would no longer dominate, thus nullifying the most crucial ingredient in the above proof. All we could predict would be that G_f must breakdown to an anomaly-free subgroup.

Appendix B

Spectral Function Sum Rules

The spectral decomposition of a two-current correlation function is

$$T\langle 0|\, J_\mu{}^A(x)\, J_\nu{}^B(0)\, |0\rangle$$

$$= \int\limits_0^\infty d\mu^2 \int \frac{d^4k}{(2\pi)^4} \frac{e^{-ikx}}{k^2 - \mu^2 + i\varepsilon} \left[-\varrho^{(0)}_{AB}(\mu^2)\, k_\mu k_\nu + \varrho^{(1)}_{AB} \left(g_{\mu\nu} - \frac{k_\mu k_\nu}{k^2}\right)\right]. \tag{B.1}$$

For any asymptotically free gauge group G_s the following relation should hold for the spectral functions. These are the spectral function sum rules [55] (SFSR):

$$\int ds \left[\frac{\varrho^{(1)}}{s} + \varrho^{(0)}\right] = 0, \quad \int ds \varrho^{(1)} = 0, \quad \int ds s \varrho^{(0)} = 0. \tag{B.2}$$

The tensor I^{AB} defined as

$$I^{AB} = \int d^4x D_{\mu\nu}(x)\, T\langle 0|\, J_\mu{}^A(x)\, J_\nu{}^B(0)\, |0\rangle \tag{B.3}$$

can be expressed in terms of the spectral functions as follows:

$$I^{AB} = \frac{3}{16\pi^2} \int ds \ln\left(\frac{\Lambda^2}{s}\right) \varrho^{(1)}_{AB}(s), \tag{B.4}$$

where we have introduced an ultra-violet cutoff Λ to regulate the momentum integral. Thus the quantity I which enters into the effective potential $\Delta\varepsilon(U)$ is expressed in terms of the quantities of the form (which is proportional to the quantity M^2 defined in (2.49)).

$$\Delta^2 = \frac{1}{F^2} \int ds \ln\left(\frac{s_0}{s}\right) \varrho(s) \tag{B.5}$$

where $\varrho(s)$'s satisfy either

$$\int \frac{ds}{s}\,\varrho(s) = F^2, \quad \int ds\,\varrho(s) = 0 \tag{B.6}$$

or

$$\int \frac{ds}{s}\,\varrho(s) = 0, \quad \int ds\,\varrho(s) = 0 \tag{B.7}$$

which depends on the group theoretic structure which will be discussed in detail below. Under G_f-transformations the currents $J^{(T,X)}$ transform as the adjoint representation of G_f. The tensor I can then be expressed in terms of as many unknown constants as there are H_f-singlets in Ad $(G_f) \otimes$ Ad (G_f).

For each multiplet of broken currents in the representation D of H_f, the corresponding combination $(P_{Ad} - \varrho_D)$ obeys SFSR's, where ϱ_{Ad} is the function associated with an IRR of unbroken currents:

$$\begin{aligned} &\int \frac{ds}{s}\,[\varrho_{Ad}(s) - \varrho_D(s)] = F_D{}^2 > 0 \\ &\int ds[\varrho_{Ad}(s) - \varrho_D(s)] = 0. \end{aligned} \tag{B.8}$$

Three cases may be distinguished:

a) H_f is trivial: Then the SFSR's we satisfied by $\varrho_{D_1} - \varrho_{D_2}$ with $F^2_{D_2} - F^2_{D_1}$ on the RHS of the first sum rule.

b) If the equivalent but distinct representation D and D' of H_f occur in Ad (G_f), then the corresponding spectral function $\varrho_{DD'}$ obeys the SFSR's:

$$\begin{aligned} &\int \frac{ds}{s}\,\varrho_{DD'}(s) = F^2_{DD'} \\ &\int ds\varrho_{DD'}(s) = 0. \end{aligned} \tag{B.9}$$

In general (NGB multiplets of) the D and D' can mix, and $F^2_{DD'}$ need not vanish. However D and D' can always be choosen such that there is no mixing, and $F^2_{DD'} = 0$.

c) If H_f is not simple, then we have

$$\begin{aligned} &\int \frac{ds}{s}\,(\varrho_{Ad} - \varrho_{Ad'}) = 0 \\ &\int ds(p_{Ad} - \varrho_{Ad'}) = 0 \end{aligned} \tag{B.10}$$

where ϱ_{Ad} and $\varrho_{Ad'}$ are spectral functions associated with unbroken currents in different factors of H_f.

Examples:

i) For the $SU(N)$ case $(N \neq 2)$

$$\text{Ad}\,(G_f) \otimes \text{Ad}\,(G_f) = 1 + \text{Ad}\,(G_f) + \text{Ad}\,(G_f) \quad \text{others}.$$

Thus there exists an SFSR associated with each H_f-invariant linear combination of the current products which does not contain the G_f-singlet. The number of independent SFSR's is one less than the number of H_f-singlets in Ad $(G_f) \otimes$ Ad (G_f).

ii) For arbitrary G_f and $H_f \subset G_f$, the total number of independent SFSR's is the number of/singlets minus the number of G_f-singlets contained in Ad $(G_f) \otimes$ Ad (G_f).

For asymptotically-free gauge theories we can assume that the spectral integrals are always rapidly convergent, so that we can saturate them with the lowest lying vector bosons of the right quantum numbers (two resonance approximation).
For the class (B.6), we get $\Delta^2 = CM_1{}^2$ where M_1 is the mass of the lightest vector resonance coupled to the currents associated with ϱ PRESKILL [12] produces an approximate numerical value for the constant C also: $C = 2 \ln 2$.
For the class (B.7), in the two-resonance approximation, we find $\varrho = 0$, and this $\Delta^2 = 0$.

Appendix C

Second Order Casimir Invariants of $SU(n)$ Groups and Related Topics

In general we can denote the most general representation of the $SU(n)$ group ($n \geqq 2$) by a Young Tableau of k rows ($k \leqq n - 1$), and j columns each row containing μ_i squares, where $\mu_i \geqq \mu_{i-1}$.

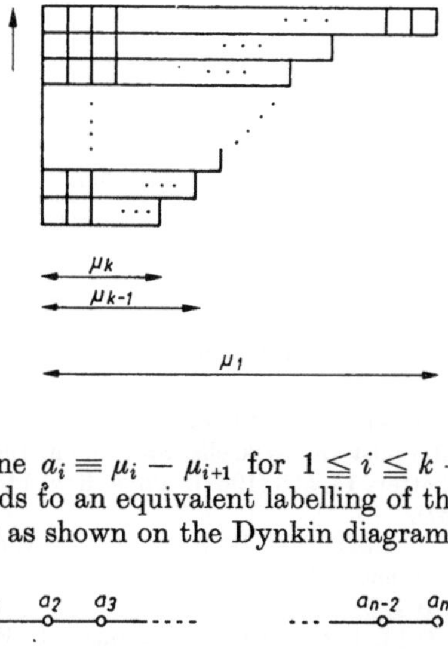

Equivalently we can define $a_i \equiv \mu_i - \mu_{i+1}$ for $1 \leqq i \leqq k - 1$, and $a_k \equiv \mu_k$. Clearly $a_i \geqq 0$; and this corresponds to an equivalent labelling of the representation according to the weights of the roots as shown on the Dynkin diagram for $SU(n)$:

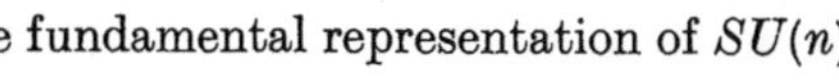

Then the second order Casimir invariant corresponding to this representation is given by:

$$C_2(\mu_1, \ldots, \mu_k) = -\frac{s^2}{2n} + \frac{n+1}{2} s + \frac{1}{2} \sum_{j=1}^{k} \mu_j(\mu_j - 2j) \tag{C.1}$$

where, $s \equiv \sum_{i=1}^{k} \mu_i$, and the fundamental representation of $SU(n)$ is normalized such that $\mathrm{Tr}\,(T_a T_b) = 1/2\, \delta_{ab}$, where T_a are the matrices of the fundamental n dimensional representation (i.e., $a_i = \delta_{i1}$ for the fundamental representation).
There are similar general formulae for the second order Casimir invariants of all the representations of the other Lie Groups[24]).

[24]) We refer the reader to a textbook by WYBOURNE [56] and the references cited therein, with a warning that the normalization used by Wybourne is not the standard normalization used in particle physics. For example $\mathrm{Tr}\,(T_a T_b) = (1/2n)\, \delta_{ab}$ for the fundamental representation of $SU(n)$ Wybourne's normalization convention.

Here we also would like to quote the following useful formulas valid not only for $SU(n)$ but for any compact simple Lie group: (Repeated indices are summed unless otherwise stated)

$$\mathrm{Tr}\,(D_a D_b) = T(D)_{ab} \tag{C.2}$$

$$D_a D_a = C_2(D)\,. \tag{C.3}$$

From (C.2) and (C.3) we can immediately obtain

$$C_2(D) = \frac{N}{\Lambda(D)}\, T(D) \tag{C.4}$$

where N is the order of the group (the number of generators) and $\Lambda(D)$ is the dimension of the representation D. Note that nothing is assumed about the reducibility of D.

For direct products of representations we have trivially

$$T(D \otimes D') = \Lambda(D)\, T(D') + \Lambda(D')\, T(D)$$

$$C_2(D \otimes D') = C_2(D) + C_2(D')\,.$$

Similarly for the direct sums of representations we have

$$T(D \oplus D') = T(D) + T(D')$$

$$C_2(D \oplus D') = \frac{\Lambda(D)\, C_2(D) + \Lambda(D')\, C_2(D')}{\Lambda(D) + \Lambda(D')}\,.$$

References

[1] S. Weinberg, Phys. Rev. Letters **19**, 1264 (1967); A. Salam, in Elementary Particle Theory, edited by N. Svarthold, Almquist and Wiksells, Stockholm, 1968.
[2] P. N. Higgs, Phys. Rev. Letters **13**, 508 (1964); Phys. Rev. **145**, 1156 (1966); T. W. Kibble, Phys. Rev. **155**, 1554 (1967).
[3] J..E. Kim, P. Langacker, M. Levine, H. H. Williams, Univ. of Pennsylvania Report Number UPR-158 T, and the references therein.
[4] S. Weinberg, Phys. Rev. **D 7**, 2887 (1973).
[5] E. Gildner, Phys. Rev. **D 13**, 1025 (1976); E. Gildner, S. Weinberg, ibid., 3333 (1976); S. Weinberg, Phys. Letters **82 B**, 387 (1979); E. Gildener, Phys. Letters **92 B**, 111 (1980).
[6] D. J. Gross, F. Wilczek, Phys. Rev. Letters **30**, 1343 (1973); D. Politzer, ibid., 1346 (1973).
[7] J. Schwinger, Phys. Rev. **125**, 1397 (1961); Phys. Rev. **128**, 2425 (1962); F. Englert, R. Brout, Phys. Rev. Letters **13**, 321 (1964); R. Jackiw, K. Johnson, Phys. Rev. **D 8**, (1973); J. Cornwall, R. Norton, ibid., 3338 (1973).
[8] S. Weinberg, Phys. Rev. **D 13**, 974 (1976).
[9] L. Susskind, Phys. Rev. **D 20**, 2619 (1979).
[10] P. Sikivie, L. Susskind, M. Voloshin, V. Zakharov, Stanford Report Number ITP-661, 1980.
[11] R. Dashen, Phys. Rev. **D 3**, 1879 (1971).
[12] M. Peskin, Nuclear Phys. **B 175**, 197 (1980); J. Preskill, Nuclear Phys. **B 177**, 21 (1981).
[13] R. Dashen, Phys. Rev. **183**, 1245 (1969).
[14] S. Helgason, Differential Geometry and Symmetric Spaces, Academic Press, New York, 1962.
[15] K. Lane, Phys. Rev. **D 10**, 2605 (1974).
[16] S. M. Bilenky, B. Pontecorvo, Phys. Rep. **41**, 225 (1978); A. de Rujula et al., Nuclear Phys. **B 168**, 54 (1980); V. Barger et al., Phys. Letters **93 B**, 137 (1980); D. Silverman, A. Soni, Phys. Rev. Letters **46**, 467 (1981); F. Reines, H. W. Sobel, E. Pasierb, Phys. Rev. Letters **45**, 1307 (1980).

[17] S. Dimopoulos, L. Susskind, Nuclear Phys. **B 155**, 237 (1979).
[18] E. Farhi, L. Susskind, Phys. Rev. **D 20**, 3404 (1979).
[19] E. Eichten, K. Lane, Phys. Letters **90 B**, 125 (1980).
[20] S. Dimopoulos, L. Susskind, Stanford Report Number ITP-681, Nov. 1980.
[21] M. Goldhaber, P. Langacker, R. Slansky, Los Alamos Report Number LA-UR-80-356, 1980.
[22] F. Reines, J. Schultz, Surveys in High Energy Physics **1**, 89 (1980).
[23] T. W. Donnelly et al., Phys. Rev. **D 18**, 1607 (1978).
[24] S. L. Adler, Phys. Rev. **117**, 2426 (1969); J. S. Bell, R. Jackiw, Nuovo Cimento **60 A**, 47 (1969).
[25] R. Peccei, H. Quinn, Phys. Rev. Letters **38**, 1440 (1977); S. Weinberg, Phys. Rev. Letters **40**, 223 (1978); F. Wilczek, ibid., 279 (1978).
[26] S. Dimopoulos, S. Raby, P. Sikivie, Nuclear Phys. **B 176**, 449 (1980).
[27] E. Eichten, K. Lane, J. Preskill, Phys. Rev. Letters **45**, 225 (1980).
[28] K. Lane, DOE Report Number DOE/ER/01545-288.
[29] S. Dimopoulos, J. Ellis, CERN Report Number Ref. TH. 2949-CERN, Sept. 1980.
[30] E. Cohen, Weizmann Report Number WIS 2/81-Jan/Ph, Jan. 1981.
[31] S. Raby, S. Dimopoulos, L. Susskind, Stanford Report Number ITP-653, Dec. 1979.
[32] H. Georgi, S. L. Glashow, Phys. Rev. **D 6**, 429 (1972).
[33] S. Okubo, Phys. Rev. **D 16**, 3528 (1977).
[34] F. Alexander Bais, J.-M. Frere, Phys. Letters **98 B**, 431 (1981).
[35] J. S. Bjorken, Ann. Phys. (N.Y.) **24**, 174 (1963).
[36] H. Georgi, Nuclear Phys. **B 156**, 126 (1979).
[37] Ling-Fong Li, Phys. Rev. **D 9**, 1723 (1974).
[38] V. Elias, S. Eliezer, A. R. Swift, Phys. Rev. **D 12**, 3356 (1975).
[39] D. Amati, M. A. Virasoro, CERN Report Number Ref. TH. 2989-CERN, Nov. 1980.
[40] C. L. Ong, U. of California at San Diego Report Number, UCSD-10P10-218.
[41] B. Holdom, Harvard Report Number HUTP-80-A010 (Revised).
[42] J. Ellis, P. Sikivie, CERN Report Number Ref. TH. 3034-CERN, Feb. 1981.
[43] S. Dimopoulos, Nuclear Phys. **B 168**, 69 (1980).
[44] K. Lane, M. E. Peskin, Nordita Preprint, NORDITA, 80/33, 1980.
[45] B. Holdom, Harvard Report Number HUTP-81-A004.
[46] L. E. Ibanez, Oxford Report Number, 75/80, 1980.
[47] M. A. B. Beg, DOE Report Number DOE/EY/2232B-220, Dec. 1980.
[48] J. Ellis, D. V. Nanopoulos, P. Sikivie, CER NReport Number Ref. TH. 3030—CERN, Feb. 1981.
[49] S. Chadha, M. E. Peskin, CERN Report Number Ref. TH. 3038-CERN, Feb. 1981.
[50] J. A. Grifols, Univ. of Barcelona Report Number UAB-FT-69, Feb. 1981.
[51] S. Coleman, E. Witten, Phys. Rev. Letters **45**, 100 (1980).
[52] G. 't Hooft, Nuclear Phys. **B 72**, 461 (1974); Nuclear Phys. **B 75**, 461 (1974), E. Witten, Nuclear Phys. **B 160**, 57 (1979).
[53] A. Dolgov, V. Zakharov, Nuclear Phys. **B 27**, 525 (1971).
[54] E. Witten, Nuclear Phys. **B 156**, 269 (1979); G. Veneziano, **B 159**, 213 (1979).
[55] S. Weinberg, Phys. Rev. Letters **18**, 507 (1967); K. Wilson, Phys. Rev. **179**, 1499 (1969); C. Bernard, A. Duncan, J. Lo Secco, S. Weinberg, Phys. Rev. **D 12**, 792 (1975).
[56] B. Wybourne, Classical Groups for Physicists, Wiley-Interscience, New York 1974.

INHALT

Die „FORTSCHRITTE DER PHYSIK" dienen der Publikation zusammenfassender Berichte über aktuelle Teilgebiete der Physik. Neben Originalberichten in deutscher oder englischer Sprache werden deutsche Übersetzungen wichtiger fremdsprachiger Berichte abgedruckt.

Aus dem Inhalt der folgenden Hefte:

H. Kresse: Dielectric Behaviour of Liquid Crystals

S. M. Bilenky, S. I. Bilenkaya, G. Motz: Neutral Current and P-Odd Effects in Processes with Participation of Leptons

I. Bender, H. C. Dosch: N-N Potentials in QCD-Motivated Quark Models

K. H. Finken: Untersuchungen an dichten z-Pinch-Plasmen

F. M. Lev: On a Three-Body Problem in Relativistic Quantum Mechanics

S. K. Bose, D. N. Tripathy: Studies of Coupled Anharmonic Oscillator Problem Using Coherent States and Path Integral Approaches

H. Stöckel: Linear and Nonlinear Generalization

Manuskripte werden nach Vereinbarung mit dem Autor von der Schriftleitung entgegengenommen.

Bei Fragen zur Produktsicherheit wenden Sie sich bitte an:
If you have any questions regarding product safety,
please contact:

Walter de Gruyter GmbH
Genthiner Straße 13
10785 Berlin
productsafety@degruyterbrill.com